I0072949

MASTERING
MODERN
SECURITY

A Smarter
Approach to
Resilient Building
Security Systems
and Staying Ahead
of Threats

ROB
JACKSON

A AUTHOR.INC

Mastering Modern Security
Copyright © 2025 by Rob Jackson
All rights reserved.

No part of this book may be reproduced, stored in a retrieval system, or transmitted in any form or by any means—electronic, mechanical, photocopying, recording, or otherwise—without prior written permission of the publisher, except in the case of brief quotations embodied in critical articles and reviews.
Published by Author.Inc
ISBN: 978-1-966372-00-4

Edited by Lisa Caskey
Cover and Interior design by Erin Tyler
Author Photo by Zach Feydo

This book is intended to provide accurate information regarding the subject matter covered. Every effort has been made to ensure accuracy; however, it is provided with the understanding that the author and publisher are not engaged in rendering legal, financial, medical, or other professional services. If expert assistance is required, the services of a competent professional should be sought.

The author and publisher make no guarantee of results from applying the ideas and information in this book. The author and publisher expressly disclaim liability for any adverse effects arising from the use or application of the information contained herein.

Printed in the United States of America
First Edition: 2025

For more information, visit: ipsystems.tech

For the employees of SafeEdge Solutions who passionately protect the customers we serve. Proud to work alongside you.

AUTHOR'S NOTE

This book is packed with insights, real-world examples, and lessons learned from the field of modern security. Some stories are based on true events, but details have been adjusted to respect privacy and confidentiality.

Your specific results will vary depending on your system setup and maintenance. The case studies and examples shared here are meant to illustrate possibilities, not guarantees. So please, don't take them as promises of specific outcomes.

When it comes to companies, all references are based on publicly available information, and any opinions shared are my own. I've done my best to be accurate, but mistakes happen—so no liability is accepted for errors or omissions.

Any endorsements or testimonials reflect the unique experiences of those individuals, and may not represent typical results. Where necessary, any relationships or connections have been disclosed.

Lastly, this book isn't a substitute for professional advice—so always consult an expert when making big decisions, especially in regulated industries. If you'd like to consult with our team, you can find us at ipsystems.tech or call us at (330) 963-0064.

Thanks for reading, and I hope these strategies inspire you to level up your security game!

CONTENTS

PART I: FACE THE HIDDEN THREATS

PART II: MAKE *S.A.F.E* THE STANDARD

PART III: TURN SECURITY INTO STRATEGY

INTRODUCTION: WHY MODERN SECURITY?

Several years ago, I witnessed a double homicide.

I was driving home from the security company I had bought just months earlier. On my way home, I stopped at an intersection and watched as a large vehicle pulled up beside another car in front of me. Suddenly, a round of gunshots went off. I was terrified.

After I watched the shooter drive off and ensured I was safe, my attention went to helping the people in the car in front of me. It was horrible getting out and walking over to the vehicle. I saw that two people had just been killed.

I called 911, gave them my name and information, and waited for the police to show up. Eventually, those 911 recordings were released to the public, exposing my personal information. I started receiving calls from news agencies and people who wanted to know what happened. I became worried about the safety of my family, as it was too easy to track us down.

I tried to resume living my life and go back to running the security business I had just purchased. Returning to work after witnessing this was challenging. The trauma lingered - I experienced flashbacks and anxiety, especially when driving. Immersing myself in work provided some distraction, but the weight of what I'd seen remained.

I'm fortunate for two key reasons. First, the working security cameras provided crucial evidence to identify and convict the shooter. This meant justice was served swiftly, reducing any potential threat to me or my family as a witness. Second, it validated the vital importance of functional security systems in real-world scenarios. As a security business owner, this incident highlighted the life-saving potential of the work we do. It wasn't just theoretical anymore. I saw firsthand how a properly maintained system could make the difference in solving a violent crime. This experience shaped my approach to security, emphasizing the need for reliable, always-functioning systems that can deliver when it matters most.

Running the security business took on new meaning and urgency. It wasn't easy, but focusing on protecting others through my business became a way to process the experience and find purpose.

I've stressed and toiled while working long hours, owning and running a security business because I believe the security systems you install in your business and home matter. They protect the assets and people you care about.

THE BURDEN OF SECURITY RESPONSIBILITY

I can imagine how worried you are about the burden of the responsibility to oversee security at your business. Your organization has asked you to check that the security systems are working. They must meet industry standards, government regulations, and your company's policies. This includes following rules from bodies like OSHA for workplace safety, HIPAA for healthcare data protection, and PCI DSS for financial security. You are also responsible for making sure the systems comply with local and federal laws on surveillance and access control. On top of all that, you need to meet the internal standards set by your own organization for protecting employees, customers, and assets. Despite all your worry and effort, systems are degrading and becoming outdated. Threats are emerging to take them offline and put your assets at risk.

I've seen the burden that outdated security systems in your business present. You log in to the camera system daily to review images and see if the camera system is recording. You ask your peers, other business executives, to approve an upgrade for a system that just went offline and is critically out of date. I've watched people like you log into their system to try to extract who's accessed the building and at what times. I've seen the stress of regulatory compliance presentations and audits, and I know you're trying your best to rely on the information that the outdated security system can provide.

But the information is inadequate, and you're worried it will lead to a security breach.

And you should be.

It's important to acknowledge that the frequency and impact of security breaches are even *higher* than what's publicly reported. Yes, news of security breaches hit the airwaves daily, but many companies opt not to disclose breaches unless legally required, especially if they can replace or mitigate stolen assets without customers finding out.

This underreporting makes it difficult to gauge the full scale of the problem. It's exponentially larger than we realize. Publicly traded companies may offer some insights through SEC filings on shrinkage, lawsuits, fines, and damages, but those reports represent just the tip of the iceberg.

Security executives carry a heavy burden of responsibility. Updating a system is hard enough. Maintaining it is something else altogether.

INTRODUCING MODERN SECURITY

This is where *modern* security comes in. Imagine a new approach to update and maintain your security system with clear standards defined. It also utilizes experts in the security field who support you in the implementation of the system and all ongoing maintenance and support. They are partners who help you update and ensure all systems are online. They use modern tools, so when it comes time to replace the aging system that's reaching the end of its life, it's already a part of the ongoing program.

By implementing the modern security system, your day-to-day reality can be much different…

You don't have to feel anxious about knowing if your camera system is working, because it's continuously monitored.

You already had the replacement budgeted and installed well in advance of the system crashing, so you don't have the burden of convincing other executives to update an offline system. It's not a fire drill anymore; it's just business as usual.

You can rest your eyes, no longer looking at poor, grainy images, because you have new cameras. Instead of logging into a computer to access data, you can use your Active Directory login and easily access all your systems from your laptop.

You don't feel alone because you have a committee of other departments supporting you in the security venture, and you have a security expert advising you on standards and industry trends.

Beyond that, you have the peace of mind that your system is compliant with regulations.

This book, *Mastering Modern Security*, will guide you through steps to successfully implement this modern, reliable system.

But first, we'll talk about the hidden dangers that exist in having an obsolete system and why the stakes are high if you allow your system to become obsolete. Then, we'll discuss the benefits of forming a cross-functional team of different departments to support your security system.

Next, we'll talk about the SAFE framework, a toolset for implementing this modern security system.

S IS FOR *SURVEY*

A IS FOR *APPLY*

F IS FOR *FORTIFY*

E FOR *EVOLVE*

Once we cover the implementation of this modern system, we'll talk about the hidden data that exists within your security system and how you can extract it to provide value to your business.

Lastly, we'll discuss a shift in mindset: moving from the traditional capital expenditure approach of buying a new system to adopting an operating expense model—just like much of the IT industry has already done.

DEEP TECHNICAL EXPERIENCE AND A STRATEGIC MINDSET

Who am I, and why should you listen to me when it comes to security?

My path to becoming an expert in security is grounded in deep technical experience and a strategic mindset. I started my career as an engineer at Boeing, working on cutting-edge projects like the 787 Dreamliner. That role took me to Seattle and Japan, where I helped launch the first of those planes. This process taught me how to blend precision, innovation, and complex systems thinking.

From there, I earned a master's degree in business and systems engineering at MIT, where I focused on how advanced technology could drive efficiency. My thesis work with Amazon gave me an inside look at using data and automation to solve real-world challenges—skills that have direct applications in today's security landscape.

After MIT, I joined McKinsey & Company as a consultant, where I advised C-suite executives on operational improvements. I saw up close how risk management is a key focus area for a board of directors. Working at McKinsey provided a world-class education in structured, strategic thinking.

This blend of hands-on engineering, cutting-edge tech insights, and strategic advisory experience has given me a unique perspective on security. I understand what it takes to design, implement, and continually improve systems that protect critical assets and people. That's why I'm passionate about helping businesses move beyond outdated models and embrace a more proactive, resilient approach to security.

All of my experiences helped build a foundation of who I am and how I think. When I bought Integrated Precision Systems (IPS) in October of 2017, I brought those to the table. I wanted the chance to own a small business and have an impact on an organization. Small business ownership was an opportunity to do that, and I loved it. I loved the exploration of technology and innovations, the executive challenges of running a business, and rubbing shoulders with other executives who have the challenge of implementing security systems in their organization.

Being an outsider in the security industry enabled me to challenge the traditional approach. I brought fresh perspectives from other industries and business settings, asked critical questions about why the traditional security approach operates the way it does, and challenged traditions if a better approach exists. I chose to write this book because what I've learned over the last seven years can benefit other security executives. This book will challenge the way they implement security in their organizations and help them implement a modern security system.

Since buying IPS in 2017, we've developed a product that monitors the health of 20,000 security devices to ensure that they're online and working. I'm proud that our business has grown to become a Top 40 Security Systems Integrator in the United States. (There are over 5,000 US security systems integrators.)

We've expanded into 30 states and 17 countries. We've replaced hundreds of outdated security systems and presented a new way to ensure they never become obsolete.

With this book, I'm sharing everything I've learned from my unconventional journey and hands-on experience in the security industry. My background in engineering, business strategy, and consulting, combined with the lessons I've gathered as the owner of Integrated Precision Systems, has given me a fresh perspective on what it takes to build resilient, modern security systems. My aim is to help other security executives rethink their approach—adopting strategies that ensure their systems stay relevant, effective, and future-proof. Whether it's through innovative technology or challenging outdated models, a smarter approach to security can make a world of difference.

SECURITY IS CRITICAL TO SUCCESSFUL BUSINESS OPS

Remember the phone book?

It still technically exists. In some neighborhoods, they still deliver it to your doorstep—a thick, clunky relic of a different era. While it used to be essential, today it's mostly a punchline. No one reaches for the phone book anymore. We've moved on to better, faster, more reliable ways to get information and stay connected.

The physical security industry reminds me a lot of the phone book. It still works. It's still out there... barely. But in a world where technology is moving fast and where threats are getting more complex by the day, relying on outdated systems is like flipping through a paper directory in a world of instant search. It's slow. It's inefficient. And most importantly—it's risky.

To truly protect your organization, security systems must be treated with the same care and strategy as IT systems. They should be planned, budgeted, and continuously maintained. Adopting a modern, proactive approach ensures that your security systems remain effective and never become a liability. Ignoring this shift can leave your organization exposed to risks like theft and data breaches, undermining all your hard work.

This book is about helping you make the shift from phone book thinking to modern, proactive security. It will help you shift from a set-it-and-forget-it model to one that is dynamic, monitored, and designed to grow with your organization.

As we embark on this journey, we'll uncover the vulnerabilities that outdated systems bring and how to tackle them, starting with

"Chapter 1: The Hidden Dangers of Your Aging Security System." There's a lot of information packed into these pages, so take breaks if you need them and jot down questions as they come to mind. This is your chance to reshape your security strategy. Let's ensure it's built to last.

CHAPTER 1:
THE DANGERS OF YOUR AGING SECURITY SYSTEM

had an interesting experience in one of the first jails I visited.

I was escorted in and had to go through a series of metal detectors, probably more than you've experienced at the airport. I had to take off my belt and give them my phone, any sharp objects I was carrying, and anything else they didn't want going inside the premises. I had to put everything into a locker. Then, I was ready to head in.

This jail had over 2,000 inmates inside, all awaiting their court date or sentencing. Some of them were charged with very serious crimes, including murder. It was more than enough to keep my head on a swivel.

I felt relatively comfortable because I was in between two very large officers, who protected my front and back. They were taking me to the control room, the brains that centrally controlled the jail.

When we got to the control room, I expected to see something modern, with a lot of new technology and tools. To my surprise, I looked up and saw an old cathode-ray tube TV (it's a big, boxy TV that we all used before the invention of flat-screens) that was showing grainy, blurry, and hard-to-see video images.

How could this jail, entrusted with keeping these inmates in the right place, keeping the officers safe, keeping the public outside safe, have such outdated technology? The cathode-ray TV was just the start. As I looked closer, I noticed the cameras were clearly antiquated models - bulky, with low resolution, and lacking modern features like digital zoom or night vision. The footage they displayed was grainy and unclear, a far cry from the crisp images modern security cameras can provide. It was shocking to see such old equipment being relied upon in a high-stakes environment like this. The contrast between the critical nature of the jail's security needs and the age of their surveillance system was stark and concerning.

I walked over to a touchscreen PC that controls the doors, and I saw that it was running Windows XP. It's been 20 years since that operating system was released, and there are well-known vulnerabilities. Yet, it was controlling all the critical doors in the jail, including the entry and exit doors.

I also discovered that the operators reported the intercoms are unreliable. Sometimes they push the button and can't hear the person talking, or the button doesn't work. As a result, they have to rely on unclear camera views to try to determine if that person is safe to open the door for.

The jail was vulnerable to an attack, which made me think: *how do we deploy modern and reliable security systems in critical-regulated spaces?*

TRAGEDY AT THE MILWAUKEE COUNTY JAIL

The stakes are higher than almost anywhere else for jails. Administrators bear the immense responsibility of ensuring the safety and wellbeing of both inmates and the officers who oversee them. It's not just about maintaining order—it's about protecting human lives in an environment where the potential for harm is ever-present.

In conversations with jail administrators, one theme always comes through: their primary concern is preserving life. Yet, gaps in surveillance, faulty locks, or neglected security measures can create vulnerabilities that escalate into dangerous or even fatal incidents. These risks aren't just operational failures—they touch on the fundamental human right to safety, even within a correctional facility.

Take the tragic 2017 case at Milwaukee County Jail, where a breakdown in security systems contributed to the death of an inmate from dehydration in a high-risk mental-health isolation area. This loss was preventable. Had there been adequate monitoring, reliable equipment, and consistent observation, that inmate's life might have been saved. This case is a painful reminder of what's at stake when safeguards fail: not just lawsuits or settlements, but irreversible human tragedy.

Facilities that prioritize safety take proactive steps, such as implementing robust inmate tracking systems, ensuring high-quality video surveillance, and conducting regular observation rounds. These measures are not just checkboxes—they are lifesaving practices that catch risks before they spiral out of control.

While financial consequences like Milwaukee County's $6.75 million settlement highlight the steep cost of neglect, the true weight of these failures lies in the emotional toll on families, staff, and communities. Investing in modern, reliable security systems isn't just about protecting budgets—it's about protecting lives. It's a moral obligation.

The lessons extend far beyond jails. Whether it's a correctional facility, a business, or a public space, the duty to safeguard people remains the same. Prioritizing safety isn't just about avoiding financial repercussions—it's about ensuring that everyone under your care makes it home safely at the end of the day.

In this chapter, we will unveil the hidden dangers of outdated security systems and expose the vulnerabilities they pose.

8 HIDDEN RISKS OF OUTDATED SECURITY SYSTEMS	
RISK TYPE	**DESCRIPTION**
RISK 1: FINES AND LAWSUITS	Financial damages caused by a failure to protect people, products, and information.
RISK 2: DAMAGING THE COMPANY'S REPUTATION	Losses to the company's brand value, as a result of security system issues.
RISK 3: INTERNAL THREATS	Damages caused by a trusted insider who seeks to do harm to the business.
RISK 4: CYBERCRIME	Unauthorized access to the security system, due to cybersecurity vulnerabilities, resulting in a physical breach, theft, and/or data breach.
RISK 5: EXPOSING TRADE SECRETS	Release of proprietary or confidential company information.
RISK 6: LOSS OF PHYSICAL AND DIGITAL ASSETS	Theft of physical property and assets from security vulnerabilities.
RISK 7: EXTENDED DOWNTIME	Loss of company revenue and customer satisfaction because the building is unable to operate/provide services.
RISK 8: LESS RELIABLE DEVICES	Poor performance of systems because of aged or unreliable equipment.

RISK, RISK, AND MORE RISK

During my time at McKinsey & Company as a management consultant, I frequently presented to chief executives. This experience helped me gain a deep appreciation for the risks they face. In particular, chief financial officers (CFOs) excel at assessing and quantifying these risks, often translating them into actionable insights for their organizations.

As a security executive, one of your critical responsibilities is to communicate the risks posed by an aging security system to your leadership team. You must assess, approximate, and clearly convey these risks to ensure that reduction plans can be put into place. The stakes are high, and proactive planning is essential to safeguard your organization.

In this section, we'll explore some of the specific risks outdated security systems pose and why they represent serious threats to your business. The risks can vary based on your industry, the services you provide, and the regulatory controls your organization must follow.

RISK 1: FINES AND LAWSUITS

There are serious risks of fines and lawsuits for failing to implement effective security systems. In 2017, Equifax was hit with a $700 million settlement for a data breach that affected 147 million people. The breach was partly due to outdated physical security systems.

This isn't an isolated incident. Companies in regulated industries face even steeper consequences. A hospital in California was fined $250,000 for failing to protect patient data, violating HIPAA regulations. Beyond fines, there's the risk of class-action lawsuits from affected individuals. If a business doesn't take physical security seriously, there is a risk of losing their permit to operate.

For example, casinos have lost gaming licenses due to inadequate security measures, resulting in millions in lost revenue and shareholder value. We've all seen the movie *Ocean's 11*, with its high-security, advanced casino, and Daniel Ocean's crew trying to get into the vault...

One day, I got a call from a casino manager. His insurance company had purchased a very inexpensive proximity-badge cloning device, one that can fit inside a backpack, to perform a penetration test to see if the vault could be accessed. The device scans the card number and other information on the badge within three feet of an employee with a valid badge, so an exact copy can be made.

The insurance company made a fake badge and had somebody dress up to look like they were supposed to be there. This person then got into the back of the house, moved around where the money is kept, and left. After that, the person submitted an insurance finding, reporting that the facility was not secure and that somebody could easily gain access into the back of the house.

Criminals are becoming more savvy, which is why the stakes are higher for companies to protect their assets.

The message is clear: neglecting security can cost your business dearly, both financially and operationally.

RISK 2: DAMAGING THE COMPANY'S REPUTATION

In 2017, the "Big Bitcoin Heist" in Iceland saw thieves steal $2 million worth of Bitcoin mining equipment from a data center, exposing glaring physical security gaps. Critical entry points lacked video surveillance, access control relied on weak single-factor authentication, and the alarm system failed to detect the intrusion in time.

Now, consider your own facility. Where is your data housed? Are your systems prepared to delay, detect, and respond to potential breaches? The stakes are high when critical information is at risk.

After visiting data centers that manage sensitive data for social media, banking, and AI operations, I've seen the difference robust security measures make. These facilities use multi-factor authentication, such as retinal scans and badge access, to ensure only authorized personnel gain entry—protecting not just infrastructure but the lives and trust of their customers.

RISK 3: INTERNAL THREATS

You're not just guarding against outsiders. Internal threats can be just as dangerous. Disgruntled employees or those with access to sensitive information can cause significant damage. Insider data theft and intellectual property leaks have led to major financial losses for companies in many industries.

Physical security is also important. In some industries, there have been cases where ex-employees gained unauthorized access to facilities, posing threats to safety. While rare, incidents like these highlight the importance of securing not only digital assets but also the physical workplace.

The stakes are high. You're responsible for protecting company assets, data, and human lives. Closing security gaps is essential.

RISK 4: CYBERCRIME

Cybercrime is on a steep rise, creating serious challenges for businesses and individuals. By 2024, the global cost of cybercrime is expected to hit around $9.5 trillion annually, with projections pushing that number up to $10.5 trillion by 2025. This surge underscores how cyberattacks are becoming more frequent and sophisticated, hitting organizations across every industry.

Old hardware and outdated security systems can't keep up with increased cybercrime and evolving cybersecurity threats. Preparing goes beyond updating hardware. It's about implementing a comprehensive, integrated security strategy.

RISK 5: EXPOSING TRADE SECRETS

Your company's trade secrets are more vulnerable than you might realize. Losing this proprietary information can result in significant financial and competitive losses. Trade secrets are often the cornerstone of what makes a business profitable and successful, yet they remain at risk without proper safeguards.

Take the 2015 incident involving ASML, a semiconductor manufacturer with patented processes that gave it a significant market edge. Despite having a physical security system in place, the company fell victim to a breach when individuals reportedly gained unauthorized access to restricted areas, stole proprietary information, and smuggled it back to China. The stolen trade secrets were later distributed among Chinese semiconductor manufacturers, undercutting ASML's competitive advantage.

The resulting lawsuit, which awarded ASML $845 million in damages, underscored the value of these trade secrets. In response to the breach, the company enhanced its security measures, introducing features like two-factor authentication to better control access and reduce vulnerabilities.

This example serves as a stark reminder: What trade secrets does your business rely on? Where are they housed, and are you doing enough to protect them? Implementing strong, modern security measures is critical to safeguarding the information that drives your success.

RISK 6: LOSS OF PHYSICAL AND DIGITAL ASSETS

The physical and digital assets inside your building are at risk, too. They are valuable and need to be protected from people who want to break in and steal them, as well as employees, freelancers, and contractors who want to sneak them out.

I want to share a story about a client that called me after a data breach. The data breach happened after hours when an unauthorized person broke into the facility, went to the server room, connected to the network, and extracted data.

Traditionally, we think of somebody breaking in and stealing physical assets that have value, like computers and other technical equipment, or goods and merchandise. We don't think about our digital assets being at risk, too.

In this particular instance, the alarm system was outdated and didn't detect the break-in to notify authorities that this was happening. Only after the fact were they able to get video footage.

RISK 7: EXTENDED DOWNTIME

An obsolete system also presents extended downtime, because if it has a hardware failure, the time to support and replace it increases significantly. Most of the time, when you have new hardware, it has a next-business-day onsite replacement guarantee. An older operating system typically isn't supported by any manufacturers. The PCs running them aren't supported, either. This limits your options for replacing that outdated computer to things like eBay or secondhand stores that sell or trade them.

For example, if your jail system goes down and you don't have an old backup unit ready to swap in, you could face long downtime getting things back online. The longer you delay replacing outdated systems, the longer the downtime will be. This downtime can lead to issues like poor inmate monitoring, higher safety risks, and potential security breaches. You're not just risking disruptions—you're risking lives, facing legal problems, and putting your institution's reputation on the line. The financial and human costs could be devastating.

RISK 8: LESS RELIABLE DEVICES

And finally, the devices become less reliable in an outdated system.

This understanding hit home when we were asked to testify about the lack of video footage, because the camera was working inconsistently and unreliably. We stated that we had notified the client that the hardware was over 15 years old, needed to be replaced, and was inconsistent operationally. But nothing ever came of it, and sure enough, when an incident happened, video footage wasn't available.

If you wait too long to replace your security system, it won't serve its purpose. Outdated software and hardware do not provide the coverage or support you're looking for.

And speaking of waiting too long, the longer you delay upgrading outdated security systems, the greater the risks overall. Moore's Law applies here—technology advances rapidly, with modern devices offering exponentially better features and security measures.

WHAT ABOUT HAZARDOUS CHEMICALS?

If your facility handles hazardous chemicals, you need to recognize the stakes involved. These materials are essential for producing many of the goods we rely on, but they also come with significant risks—not just to your organization, but to the surrounding community and environment. Bad actors could exploit these materials, and inadequate security opens the door to disaster.

Take the explosion at Bayer CropScience in West Virginia in 2008.

Two workers lost their lives, and an investigation uncovered serious security flaws: poorly secured perimeters, weak access controls at high-risk areas, and an emergency response plan that was neither well-documented nor routinely trained on. The consequences were massive.

Families of the workers filed lawsuits. Nearby residents reported health issues like breathing problems and eye irritation. Local businesses were forced to shut down temporarily, losing revenue while safety concerns were addressed. Roads were closed for days, and the air quality made daily life unsafe. The chemical contamination extended to soil and water, requiring costly, long-term cleanup efforts and ongoing environmental monitoring.

The fallout didn't stop there. The company faced steep fines for violating safety and environmental regulations. Production delays cost them millions in lost sales and penalties for failing to deliver on time. Regulatory inspections became more frequent and stricter, consuming staff time and resources. Executives were pulled away from their core responsibilities, scrambling to manage legal battles, media attention, and damaged relationships with the local community. Rebuilding trust with neighbors, officials, and investors became a long-term challenge.

It's also important to note that some businesses have multiple risk types. In the case of Bayer CropScience:

RISK #1: FINES AND LAWSUITS
RISK #2: DAMAGING THE COMPANY REPUTATION
RISK #6: LOSS OF PHYSICAL ASSETS
RISK #7: DOWNTIME AND PRODUCTION LOSS

This incident highlights how a single security failure can spiral into widespread consequences. It's not just about lawsuits or lost revenue—it's about protecting lives, maintaining public trust, and ensuring your organization can continue operating without disruptions. When hazardous chemicals are involved, the cost of inadequate security is far greater than any upfront investment in prevention.

A MODERN SECURITY APPROACH

The traditional physical security industry is skilled at replacing outdated systems, often solving immediate issues effectively. However, the real transformation happens when organizations shift from replacing systems to maintaining and supporting them long-term. A modern approach to security includes not just installing systems but also planning for ongoing maintenance and future upgrades.

Failing to plan for long-term maintenance puts everything at risk—frequent breakdowns, costly emergency repairs, and extended vulnerabilities. Outdated systems can't keep up with evolving threats, leaving your organization exposed to compliance failures, legal issues, and security breaches. Without a proactive approach, you're stuck reacting to crises rather than preventing them.

Take the example of a jail where 30% of cameras and 60% of intercoms weren't working. Leadership recognized the risks and replaced the entire system, transforming their approach with a 15-year support plan that included scheduled equipment replacements. Today, the control room features high-definition cameras, updated touchscreens, and reliable intercoms, making operators' jobs safer and more efficient. Detailed performance reports ensure 99.9% uptime, with plans in place to quickly address any issues.

This proactive model has revolutionized how the facility operates. Ask yourself: do you have a plan to keep your security system updated and reliable for the long term? If not, now is the time to develop one.

The global interconnected economy has transformed the world, scaling operations from mom-and-pop shops to massive Walmart and Amazon distribution centers holding millions in goods. This consolidation brings benefits, but it also concentrates sensitive data, critical information, and valuable assets under one roof—making them prime targets for security threats.

Chemical facilities with proprietary technologies, data centers processing financial transactions and personal information, and correctional facilities responsible for human lives—all require vigilant protection. Security executives like you bear the responsibility of mitigating these risks and safeguarding your organization. The scale and complexity of modern security challenges demand more than individual effort.

Today's threats are more complex and sophisticated than ever, capable of breaching even robust systems. Overlooking these dangers puts businesses at unprecedented risk.

If not, you could find yourself in a whole lot of hurt. In the next chapter, we'll explore how those risks impact you, the security professional.

CHAPTER 2:
HOW RISK IMPACTS YOU

When I first entered the security industry, I expected everyone to come from a similar background—maybe military, maybe law enforcement. But I quickly discovered just how diverse this field is. I didn't come from a traditional security path myself. I studied engineering, earned a master's degree from MIT in systems engineering, and spent the early part of my career in aerospace, then I spent time at Amazon, and finally, I was a management consultant. My approach to problem-solving is grounded in systems thinking and data. Stepping into this space, I realized how much value different backgrounds bring to the table.

I've met security leaders who started in IT, others who came up through the trades—construction, facilities maintenance, even project management. Some are deeply tactical, having served in the military or law enforcement. Others are strategic planners or technology experts. Each of these backgrounds brings a different way of assessing threats, influencing leadership, and implementing solutions. No two professionals look at a risk the same way—and that's not just okay. It's essential.

In this chapter, we'll walk through five common profiles of security professionals. You'll likely see yourself—or your colleagues—in at least one of them. Understanding these types can help you build stronger teams, communicate more effectively, and grow into the kind of leader who brings out the best in others.

THE FIVE TYPES OF SECURITY MANAGEMENT PROFESSIONALS

We will discuss the various departments required to effectively manage your company's security. Security systems are the backbone of modern business operations, but behind these systems are the professionals who design, implement, and oversee them. Security leadership comes in many forms, shaped by distinct professional backgrounds, expertise, and approaches to problem-solving. To better understand how organizations can effectively leverage their security systems, it's critical to explore the types of security executives and what they bring to the table.

Throughout my years working with security executives, I found that there are five general personality or professional types typically involved in security management.

SECURITY MANAGEMENT PROFESSIONAL TYPES	
EXECUTIVE TYPE	**KEY STRENGTHS/EXPERIENCE**
IT SECURITY EXECUTIVES	Understanding of networking, software applications, and cyber security requirements.
LAW ENFORCEMENT AND MILITARY VETERANS	Able to identify vulnerabilities and countermeasures from their experience.
CORPORATE SECURITY VETERANS	Familiar with setting policies, executing a budget, and avoiding common security pitfalls.
FACILITIES AND OPERATIONS LEADERS	Supports the physical hardware and repair issues that arise. Practical and able to ensure the systems support efficiency.
RISK AND COMPLIANCE PROFESSIONALS	Able to identify and quantify risks that the business faces. Uses the security system as a tool to reduce risk.

1. IT SECURITY EXECUTIVES

The first type is an IT security executive. These security professionals typically come from backgrounds in networking, systems administration, or cybersecurity. They excel in the technical aspects of security systems, with a strong focus on system integration, software, and cloud-based solutions. IT security executives are invaluable for ensuring that digital and cyber defenses are airtight.

While their technical expertise is unparalleled, they may struggle with the physical security aspects of their role, such as building access control or surveillance. Bridging the gap between the virtual and physical worlds often requires collaboration with other departments.

IT security executives shine when maintaining, integrating, and upgrading complex security infrastructures. They are often the first to advocate for scalable, cloud-based platforms that can adapt to evolving threats.

2. LAW ENFORCEMENT AND MILITARY VETERANS

The second type of security professional is a law enforcement or military veteran. They bring a wealth of experience in threat detection, risk assessment, and emergency response. They excel in creating and enforcing procedures, training programs, and partnerships with local authorities.

Law enforcement and military veterans' expertise often lies in traditional security measures. This may leave them less comfortable with emerging technologies and system integration. They may need support in navigating rapidly evolving tech landscapes.

These leaders are critical for high-risk environments that require strong emergency preparedness and disciplined security procedures. Their ability to instill a culture of vigilance and readiness is invaluable.

3. CORPORATE SECURITY VETERANS

Next up are *corporate* security veterans. These professionals are adept at blending physical and digital security knowledge. They understand corporate dynamics and excel in building security programs that align with business objectives. Metrics, ROI, and business case development are where they shine.

While they may lack the deep technical expertise required for some cutting-edge security technologies, they are strong in strategy and alignment with corporate goals.

Corporate security veterans thrive when tasked with developing comprehensive security programs from the ground up, navigating corporate politics, and ensuring that security is seen as a business enabler, not a cost center.

4. FACILITIES AND OPERATIONS LEADERS

Facilities and operations know the ins and outs of building systems. They are skilled in coordinating with contractors, managing budgets, and ensuring practical implementation of security measures.

Their expertise is typically centered around logistics and project management. Without the help of a robust team, this may leave them less prepared to handle overarching security strategies or regulatory complexities.

These leaders are instrumental in the practical implementation and maintenance of physical security systems, ensuring they work seamlessly with other building infrastructure.

5. RISK AND COMPLIANCE PROFESSIONALS

Risk and compliance experts bring deep knowledge of regulatory requirements and standards and excel at creating documentation, preparing for audits, and ensuring adherence to policies and procedures.

While governance, risk, and compliance experts are meticulous, they may lack the hands-on experience needed for designing or implementing physical and technical security solutions.

Risk and compliance leaders ensure that security measures are effective and legally compliant, protecting organizations from fines, lawsuits, and reputational damage.

As you can see, there is no one-size-fits-all security executive. Each type of leader brings unique strengths and perspectives to the table.

THE UNIVERSAL BURDEN OF SECURITY RISKS

Every type of security professional has their own unique strengths and weaknesses, but the challenges and pressures they face are strikingly similar. Regardless of their path to the position, security executives share common vulnerabilities in managing security risks.

CAREER DAMAGE FROM MAJOR BREACHES

The professional fallout from a significant security incident can be devastating. For security executives, their career often hinges on their ability to protect assets and mitigate risks. A single failure can undermine years of success, making the role one of high pressure and limited room for error.

I've seen it happen more than once: a security executive who wasn't even present when the breach or incident occurred still ended up losing their job. At the end of the day, the facility was their responsibility. The systems were their responsibility. And the people? Also their responsibility.

For example, a warden might lose their position after a riot or fatal incident, even if they weren't directly involved. That's because leadership accountability doesn't start and stop at your office door. It's tied to how well you've maintained the systems, trained your teams, and planned for emergencies.

When a breach happens and lives are lost—or even just put at risk, the spotlight quickly turns to leadership. In some cases, it's career-ending.

PERSONAL LIABILITY FOR FAILURES

Every security executive understands the personal stakes involved. A major failure, such as a data breach or physical security lapse, can lead to questions of personal accountability. In some cases, this liability extends beyond reputational harm to legal consequences, placing an immense burden on the individual.

LEGAL AND REGULATORY CONSEQUENCES

Failing to comply with laws, regulations, and standards—whether related to data protection, workplace safety, or surveillance—can lead to fines, lawsuits, or operational shutdowns. Security executives must ensure compliance while navigating a complex regulatory landscape.

CONSTANT STRESS OF "WHAT IF" SCENARIOS

Security leaders must live in a state of perpetual vigilance. They must always ask, "What if this system fails?" or "What if this threat materializes?" This mindset, while necessary, can be mentally exhausting, as it requires anticipating and preparing for an ever-expanding range of risks.

For most people, work ends when the office lights go off. But for security executives, it never stops.

This role isn't a typical 9-to-5 assembly line. The threats and risks to people, property, and operations don't clock out—and neither do you. There's a constant mental loop running in the background: Is the system still working? Are we taking enough proactive steps? What happens if something goes wrong?

You're the one who gets the call in the middle of the night if there's an incident. You're the one who tells your staff, "Call me anytime—doesn't matter if it's 2 a.m.—I'll pick up." And you mean it, because you know that one missed alert could lead to major consequences.

Many security executives I know subscribe to real-time alert feeds—local riots, extreme weather, public safety threats—because they need to know if something's unfolding near their facilities. One executive I spoke with lives in California and said the recent wildfires kept him glued to his phone, wondering, Are our people safe? Is the building okay? Are our evacuation plans current?

That's the stress that comes with the role: the weight of knowing that if something goes wrong, all eyes turn to you. It's not just about whether the cameras are online—it's about whether the people under your watch make it home safe.

24/7 RESPONSIBILITY

Threats don't adhere to office hours, and neither can security executives. The responsibility for ensuring systems are operational and threats are mitigated is unrelenting. Executives must remain available at all times to respond to crises, adding a layer of personal sacrifice to the role.

HIGH VISIBILITY WHEN THINGS GO WRONG

While success in security often goes unnoticed, failures are glaringly public. A breach, theft, or violent incident can draw intense scrutiny, not only from within the organization but also from the public and media. Security executives are often the ones in the spotlight, expected to provide answers and solutions under pressure.

BUDGET JUSTIFICATION PRESSURE

Securing adequate funding for security initiatives is a perennial challenge. Security executives must understand the risks and articulate them convincingly to other stakeholders, justifying investments in systems, staff, and upgrades. This requires balancing the need for protection with the fiscal realities of the organization.

One of the most surprising burdens security executives face isn't just managing threats. It's constantly having to justify the cost of preventing them.

Security leaders are in a perpetual state of selling. When budget season rolls around, they're not just filling out paperwork. They're writing persuasive cases, preparing presentations, and lobbying executives for funding. It's not enough to say, "This system is failing"—you have to prove it, show the risk, and convince a committee that the dollars should be allocated.

I think about one client of mine—a former Secret Service agent. This is someone who protected the President of the United States and his family. Now? He spends a significant portion of his time writing up budget justifications and calling business unit leaders, trying to get a 20-year-old system replaced.

When you consider someone with that level of training, experience, and skillset—having to convince a boardroom that a camera system should be upgraded—it feels out of place. And yet, it's the reality for most security professionals today.

Of course, these requests should be reviewed and questioned. But the amount of time and energy spent convincing others that outdated systems need to be replaced is time not spent on strategy, planning, or leading. It's a quiet, ongoing pressure that's hard to quantify—but every security exec knows it well.

The role of a security leader carries an inherent weight, regardless of executive type, amplified by the high-stakes nature of their responsibilities.

THE UNIQUE IMPACTS OF SECURITY RISKS

All security executives face significant pressures, but the way these risks manifest depends on their professional background. Each type of security leader approaches challenges with their own expertise and blind spots, leading to unique impacts when security risks arise. Here's a closer look at how these differences play out:

1. IT SECURITY EXECUTIVES

During a security incident, IT security executives focus on technical issues. They analyze network logs, update firewalls, and check for breaches in the digital infrastructure. Their concern revolves around metrics like uptime, vulnerability scans, and system patches.

This technical focus can lead them to overlook critical physical security elements. For instance, they may miss vulnerabilities like tailgating through access-controlled doors or unmonitored building entrances. While they excel at mitigating cyber threats, the human and physical aspects of security may not receive the same level of attention.

Coming from an IT systems and engineering background, I often approach security problems like an engineer would—through system diagrams, data flows, and network architecture. However, I don't see physical blind spots and vulnerabilities. Over time, I've learned to deeply value the experience of law enforcement and corporate security professionals, who approach things from a completely different perspective.

I remember walking through a facility with a former police officer who casually pointed out, "You see that blind spot where there's no camera coverage? That's where you'll have a problem." It was instinctive for him. He'd been to enough crime scenes to recognize the patterns: unmonitored corners, forgotten entrances, access-controlled doors that were easy to tailgate through. As someone trained in IT, those wouldn't have been my first concerns.

That's the gap. While IT security execs excel at securing data and cyber infrastructure, they can overlook the human element—like tailgating vulnerabilities or unmonitored building entrances. Without a background in physical threat environments, these risks may not register until it's too late.

IT executives often worry about malware, phishing, and data

breaches, which dominate their field of vision. However, when an incident crosses into the physical realm, their lack of experience in those areas can exacerbate risks.

2. LAW ENFORCEMENT AND MILITARY VETERANS

Security executives with law enforcement or military backgrounds excel in tactical readiness and operational security. They are focused on patrol routes, emergency response protocols, and personnel readiness when implementing measures like badge readers or access controls.

Their tactical strengths can lead to gaps in understanding IT systems and technical integration. For example, while they ensure that guards are perfectly trained and patrol schedules are flawless, they may struggle to integrate badge reader systems with IT networks or troubleshoot database connectivity issues.

I've worked with several security professionals who came out of law enforcement or military service. Their resumes are impressive— decades of experience handling crisis situations, command-level leadership, and deep knowledge of threat response. These are the folks you want beside you in a real emergency. But when those same professionals transition into corporate security roles, there's sometimes a disconnect between tactical instincts and long-term strategy.

I remember working with a former police lieutenant who took over security for a multi-site distribution company. He knew how to respond quickly when something went wrong, but he struggled to get buy-in for preventive upgrades. He insisted on replacing hardware after it failed or running drills that echoed a SWAT response. When asked to present a five-year roadmap, budget justification, or vendor evaluation framework, he felt out of his depth. His focus was rooted in reaction, not risk modeling.

Another challenge? Many of these leaders are used to tight chains of command and quick decision-making. In the corporate world, security decisions often involve collaboration—legal, finance, HR, IT—and require influencing peers without rank authority. One military veteran I met confessed that the hardest part of his role wasn't designing the security plan. It was sitting in meetings trying to convince the CFO that the expense was worth it.

Their anxiety often stems from ensuring physical and procedural readiness, such as making sure guards are positioned correctly or that evacuation plans are sound. However, compliance with technical and regulatory aspects may create additional stress.

Their training gives them courage to handle that stress. However, they may not always recognize the importance of system health monitoring, analytics dashboards, or standards-based design documentation. It's not a lack of capability—it's a gap in exposure. That's where the modern security approach comes in: helping veterans evolve from operational warriors into strategic stewards of long-term risk reduction.

3. FACILITIES AND OPERATIONS LEADERS

Facilities and operations leaders are well-versed in physical infrastructure. After an incident like a theft, they immediately focus on tangible aspects such as door hardware, camera placement, and building layout. Their expertise ensures that physical vulnerabilities are addressed efficiently.

Their focus on physical security can come at the expense of system-wide risks, particularly software vulnerabilities or the cybersecurity aspects of access control systems. For example, they might miss critical flaws in how access logs are stored or fail to notice that outdated firmware creates system vulnerabilities. Their primary concern lies in the integrity of the physical infrastructure—doors, locks, cameras—while broader risks involving data breaches or system integration may go unnoticed.

I worked with a large manufacturing facility where the operations manager prided himself on uptime. To his credit, the equipment was humming, workflows were efficient, and production was steady. But the security system? It hadn't been touched in over a decade. The cameras were so outdated they could barely make out license plates. One exterior gate badge reader had been broken for six months, and visitors were piggybacking in behind staff daily. When I asked about it, he said, "Well, it still works most of the time, and no one's broken in yet."

That "good enough" mindset might work for some maintenance issues—but not for security. Facilities leaders are often stretched thin, juggling competing priorities. Without clear standards or system monitoring, outdated tech can fly under the radar. By the time something goes wrong—whether it's a theft, safety incident, or compliance audit—it's often too late to prevent the damage.

Modern security isn't about waiting until it breaks. It's about helping these leaders adopt a proactive approach to keep things running and keep people safe.

4. CORPORATE SECURITY EXECUTIVES

Corporate security executives are skilled at balancing both physical and digital security concerns while navigating the business environment. When incidents occur, their focus shifts toward assessing the overall impact on business operations and presenting solutions in terms that align with corporate priorities, such as ROI or brand reputation.

While they have a broad understanding of security needs, their generalist approach can sometimes lack depth in specific technical or tactical areas. For example, they excel at presenting a strong business case for upgrading a security system but overlook granular details like camera resolution limitations or a software integration issue.

I've worked with corporate security executives who know exactly what's wrong with their systems, but they can't get the approvals they need to fix them. One told me, "I've asked five times for budget to replace that intercom system, and every time it gets kicked down the road." That kind of delay is frustrating when you're responsible for safety and know the consequences of failure.

Their anxiety often revolves around maintaining executive buy-in and ensuring that security is viewed as an enabler rather than a cost center. They feel the pressure to demonstrate measurable value from security investments, especially in the aftermath of an incident when questions from leadership are most pointed.

These professionals are often exceptional at managing incidents in the moment. We're helping them become just as proactive about long-term system health and modern risk mitigation.

5. RISK AND COMPLIANCE PROFESSIONALS

Risk and compliance executives excel in regulatory adherence, audit preparation, and policy enforcement. They quickly identify and address gaps in documentation, standards, or governance during a security audit.

Their policy-focused mindset may leave them less prepared for practical, real-world security challenges. For instance, they may fail to notice a camera's blind spot or underestimate the effectiveness of guard deployment strategies, focusing instead on whether protocols meet standards on paper.

I've worked with compliance leads who didn't care what brand of camera was installed. They just wanted to know, "Can we prove that it's working? Are we logging who's accessing secure areas? Are we audit-ready?" Their priority is making sure nothing creates legal exposure, and they tend to spot problems others miss—like expired credentials, shared logins, or systems that haven't been updated in years.

Their anxiety is centered on policy violations, non-compliance, and audit findings. While they ensure the organization stays on the right side of the law, their focus on documentation can lead to underestimating real-world vulnerabilities.

They bring accountability to the table. They help translate system weaknesses into regulatory and reputational risks. When you empower them with real-time data and reliable systems, they become some of your strongest allies in pushing for better security.

Each security executive type faces unique sets of challenges. Their professional background profoundly influences how they perceive and prioritize security risks. It determines which threats dominate their thoughts—and, just as importantly, which vulnerabilities they may unintentionally overlook.

The most effective security programs are led by executives who understand their strengths and proactively build teams to address their gaps. Success lies in collaboration, whether it's integrating advanced technology, responding to emerging threats, or navigating corporate expectations. Security is no longer a single department's responsibility—it requires a coordinated effort across IT, operations, facilities, and compliance teams.

The first step to overcoming challenges is understanding how your background shapes your perception. Recognize your biases, build strong teams, and adopt systems that address risks holistically. When you do, you will sleep easier—knowing that both seen and unseen threats are being managed effectively.

WHY HOLISTIC TEAMS ARE ESSENTIAL

As a security executive, you likely fit into one of the five categories we've covered. While you have distinct and valuable strengths, there are areas where additional support can be beneficial. A committee can complement your expertise by bringing in specialists from other relevant fields. By recognizing the diversity in security leadership and leveraging the strengths of each role, you can build a resilient, modern, *holistic* security system that stands the test of time.

For example, if your background is in security, you may not have the same depth of knowledge in networking, IT systems, or cybersecurity. Involving experts from these areas can strengthen your approach. Ultimately, the responsibility for the system's success rests with you, but leveraging a committee ensures you have the support needed for a smooth and effective implementation.

All security executives can thrive despite the universal pressures of the role. Security leadership is inherently demanding, but with the right tools and support, it's possible to meet those challenges head-on and safeguard both organizational assets and personal well-being. The solution isn't to become an expert in every aspect of security—it's to acknowledge where your expertise begins and ends. Each team member brings a different lens to security risks, collectively creating a comprehensive approach that minimizes blind spots. Building a multidisciplinary team that complements your background ensures all angles are covered.

No security executive can shoulder the burden alone. The weight of these shared challenges is why a modern, comprehensive security system is essential. Strong teams, built to complement the leader's strengths and fill in their gaps, are vital.

In this chapter, we covered the different types of security executives and the common challenges they face. In the next chapter, we'll cover why effectively managing a modern security system requires a team.

CHAPTER 3:
MANAGING SECURITY REQUIRES A TEAM

O ver the last couple of years, I've had the chance to work with a security executive at a manufacturing facility. This person's background was in facilities and maintenance, and they were given the responsibility by company executives to implement security. It makes sense that the security system is connected to facilities because it's a wired system inside a building and controls the doors.

One of the first things this executive did was implement a new video surveillance system in one of their buildings. It was a successful implementation, but there were some bumps along the way that IT and HR disliked.

IT was concerned about the system's cybersecurity vulnerabilities and its impact on network bandwidth. They weren't consulted early enough in the process and felt the new system didn't meet their security standards.

HR raised issues about employee privacy, particularly in areas like break rooms and near restrooms. They were worried about potential legal implications and the impact on employee morale.

Both departments also had concerns about data storage, retention policies, and who would have access to the footage.

As a result of these bumps, we began working together. I was brought in as a consultant and started working with the security executive to develop a comprehensive security plan that addressed these interdepartmental concerns.

The plan wasn't something that he or I alone could do. It was something we did collaboratively, bringing together the IT, HR, legal, and compliance departments. With those perspectives, the security executive and I were able to put together a solid security plan that balanced the needs and interests of all the different parties that interact with the organization.

In this chapter, we'll explore why effective security management in today's complex business environment requires a collaborative effort across multiple departments and functions, and how failing to adopt this approach can leave your organization vulnerable.

KEY SECURITY COMMITTEE DEPARTMENTS

Creating an effective security system requires the collaboration of key individuals *and* departments across your organization to ensure the system is comprehensive, compliant, and well-maintained. It demands more than installing equipment and working with a few security professionals. Each department brings specialized knowledge and unique perspectives that are essential to the success of the security strategy.

Let's talk about some of the key departments or individuals that you need to pull into your security committee to help you have a successful implementation:

IT

IT individuals are knowledgeable in networking, cybersecurity, user access, and data protection. They're an important voice in ensuring that the security system you implement is secure from cybersecurity threats and in compliance with the organization's IT guidelines.

LEGAL AND RISK

The general counsel or legal team within an organization is responsible for an organization's compliance and liability mitigation. They play a crucial role in ensuring that your security system meets regulatory and insurance requirements. Their input helps raise important concerns and ensures that legal and compliance standards are properly addressed.

HUMAN RESOURCES

HR is one of the gateways into a company. They know who's authorized to access a building, what their title is, and when somebody has been removed from an organization. An HR representative should be involved in managing building access. They play a key role in ensuring that anyone issued a badge or credential is properly authorized to enter. Their input helps guarantee the system meets those standards.

HR also brings a critical perspective on data privacy. For example, with the implementation of GDPR in Europe, there's heightened

awareness around protecting individuals' privacy. HR professionals are typically well-versed in these requirements. If certain
areas shouldn't have video surveillance or badging systems that
track specific activities, HR will raise those concerns and provide
valuable guidance.

FACILITIES

The security system is attached to the building it secures. Wires are
run through and powered by the facility it's protecting. Facilities
need to be involved in the physical security system for both the
deployment and the system's ongoing maintenance.

SECURITY

A security professional can develop the overall strategy of a security
system, the known threats, and what the entry points are. Having
that perspective on the team is essential.

Building a security system is a collaborative effort that requires
input from multiple departments to address all potential risks and
challenges. No single individual or department can handle every
aspect effectively. By forming a well-rounded security committee,
you gain valuable expertise from across the organization and ensure
that your system is secure, compliant, and well-maintained over
time. When every department's voice is heard, the implementation
becomes more efficient, and your organization is better protected
against potential threats.

THE BENEFITS OF A CROSS-FUNCTIONAL SECURITY COMMITTEE

Now that we've identified what departments to involve in your security committee, let's talk about some benefits of this committee. The cross-functional perspective of a security committee can provide:

COMPREHENSIVE RISK ASSESSMENTS

By bringing together the perspectives of all these different departments, you have a comprehensive view of what risks an organization faces and the severity of those risks. I've enjoyed working with security committees over the years because of the surprises and insights that come up as we identify the risks within an organization.

I remember working on a physical site layout to determine where we would place card access systems. Although I had never been to the facility, I acted as a consultant, helping them understand the system's objectives and how to implement it effectively. This facility housed biological pathogens that needed to be stored and secured in a highly controlled environment.

If I had developed a security plan without input from operations (who emphasized which areas needed to be protected) or legal and compliance (who clarified the documentation required for annual audits), we would have missed critical requirements. For example, we learned that one area required card access with two-factor authentication, restricted to a very limited list of individuals. The insights from the cross-functional team during the risk assessment were invaluable, ensuring the security plan met both operational needs and regulatory standards.

A BALANCED APPROACH

Another benefit of having a committee is that it prevents one individual or department from having an overly authoritative perspective or influence on the outcome.

A security professional may want more cameras and more card access, but having the perspective of HR, operations, and finance helps balance how far you go and what objectives you are trying to meet. The security professional has valid points—there are things that they're trying to accomplish with the additional camera coverage or card access they're implementing. Because this trade-off (and back and forth) happens in the security committee, the situation ends up with a balanced and reasonable outcome.

IMPROVED COMMUNICATION AND COORDINATION

In the opening example of this chapter, we talked about a facilities person that implemented a security system. The problem is they didn't communicate appropriately across the departments on what they were doing or how they were doing it. The result was a bumpy implementation with pushback and resistance to the outcome.

When we shifted to bring in a committee to include multiple stakeholders, we felt and saw the difference. These departments knew what was coming and accepted and contributed to the outcome, so the communication and acceptance of the implementation was very successful.

ENHANCED COMPLIANCE AND RISK MANAGEMENT

The executives within a business do not want to be surprised with fines, damages, or lawsuits. When you bring a collaborative team together, you better identify and reduce overall risk.

By bringing together a collaborative team, you ensure that you are implementing compliance and appropriately managing risk.

THE POWER OF COLLABORATION

As the security executive and I conducted an after-action report on the security plan's implementation, it became clear that we needed additional expertise. Fortunately, other departments within the organization were eager to get involved, offering support to help make the new security executive successful.

We brought in a director from IT who was particularly interested in the network configuration. He wanted to know how systems were connected to Active Directory, what cybersecurity policies were applied to the equipment, and how passwords were being managed. We also involved someone from HR who was concerned about protecting employee privacy but also wanted surveillance in key areas to address specific safety concerns. Their focus extended beyond safeguarding physical assets—they were equally invested in the wellbeing and protection of employees.

We began developing a security plan that documented the risks each department identified and outlined strategies for implementing a system to manage or mitigate those risks, bringing these stakeholders together. It took multiple meetings to align everyone's input. Working by committee is slower than having one person install a system independently. But, the outcome was far superior.

The result was a comprehensive playbook—a guide for implementing the security system. Since then, we've used this playbook for subsequent implementations and upgrades, and the process has gone smoothly. Thanks to the early collaboration where stakeholders voiced their concerns, set clear standards, and aligned on a plan, we no longer encounter the challenges experienced during the initial rollout.

This process transformed departments from skeptics or critics into champions and active supporters of the system. They understood its purpose, supported its implementation, and the result was a seamless, frictionless rollout.

HOW TO SET UP AND IMPLEMENT A SECURITY COMMITTEE

Now that we've identified what departments you want to include in your security committee and the benefits of each, we should talk about how to bring this group together.

In my experience setting up and implementing security committees, a few key considerations can make a significant difference.

STEP 1: IDENTIFY YOUR EXECUTIVE SPONSOR

First, it's essential to have an executive sponsor who can help you communicate across departments and secure the necessary resources. This top-down approach ensures buy-in from the executive team, making it more effective than trying to rally support horizontally across the organization on your own. The key question is: Who is your executive champion? Identifying that person, establishing a clear line of communication with them, and gaining their support is crucial to the success of your committee.

STEP 2: SELECT YOUR COMMITTEE INDIVIDUALS

Second, carefully select the individuals you want on your committee. While it's important to involve the right departments, you also need to identify specific people within those departments who can contribute meaningfully. Aim for a balance between implementers and strategic thinkers—people who can provide high-level guidance and are willing to roll up their sleeves to participate in making the security system a success.

In many cases, an IT executive takes on the overall responsibility for steering the security plan. However, they often delegate a director to participate in committee meetings and report back on progress, ensuring the executive remains informed without being directly involved in every discussion.

STEP 3: SET YOUR SECURITY GOALS

Now that we've covered the departments to include and the value of forming a security committee, let's talk about how to set it up for success. The goal is to establish a clear plan, have specific objectives, and run an organized process. The committee's role is to develop standards and guidelines that will ensure your security system operates effectively within your business.

A specific example of an ask I've seen security executives make is: "I would like to lead our organization through the setup of a video surveillance system standard over the next three months." This clear statement outlines both the objective and the timeline, with a defined deliverable—a set of guidelines for implementing a security system within the organization.

STEP 4: FRAME YOUR ASK

Once you've articulated this goal, the next step is to frame your ask to the executive team. You might say, "I would like to schedule a one-hour meeting with a member of each relevant department. These meetings will take place once a week for two months to help develop this plan. I'll come prepared, run the meetings, track action items, and lead discussions to identify risks. Together, we'll document the priority and severity of those risks and create a security standard to address them."

This process provides structure. You have a clear plan, a specific ask, and a framework to follow across the committee. The next critical step is securing an executive champion. Presenting this plan to an executive is essential to gain the buy-in and support needed to move forward. Start small—don't try to accomplish everything at once. Focus on setting one standard and tackling one key issue. This will build momentum, making it easier to gain traction for larger efforts over time.

One security executive I worked with used this approach to develop video surveillance standards. We outlined the objective—to have a completed document within three months—and identified the necessary stakeholders to approve the plan. From there, we created a deliverable timeline to stay on track. Once the plan was in place, we presented it to their executive team to secure buy-in and resources for implementation.

This 4-step process is a practical, straightforward guide for quickly and effectively establishing your security committee.

IMPLEMENTATION CHALLENGES

While committees can be highly effective in guiding security initiatives, they are not without challenges. It's important to acknowledge and prepare for these hurdles to ensure the process runs smoothly. Understanding these obstacles will help you manage expectations, stay patient, and set your committee up for success.

Let's explore three of the most common challenges you may encounter when working with committees.

They're slower. You might feel impatient at times and want to just get moving, but a committee's balanced approach is slower to discuss things and make decisions. Be patient—it's a better outcome.

Some individuals don't think they're needed. We all know there's a bit of a cowboy energy that we all have, and people just want to take action. But, in these important regulated industries, it is worth pausing. It's worth slowing down. You need a balanced approach; you need the committee.

Lack of executive support and direction. It is hard to get resources from other departments if an executive isn't supporting you or if it's not deemed important. Having that executive champion is crucial, and not having one is a pitfall to watch out for.

While committees come with their fair share of challenges, understanding these pitfalls will help you navigate them more effectively. With patience, clear communication, and the right executive support, you can overcome these obstacles and ensure that your committee functions as intended. A well-run committee takes more time and effort, but it ultimately leads to stronger, more sustainable outcomes for your security initiatives.

CASE STUDY: FACIAL RECOGNITION ANALYTICS

A physical security system was evaluated at a medical facility, which faced a unique challenge: they were required to provide access to anyone seeking medical care. As a public space, their doors were to remain open to everyone. However, this created a delicate balancing act, as doctors and nurses—those providing care—could be at risk from individuals entering the facility who may pose a threat.

I was fortunate enough to sit in on a hospital committee that was evaluating the use of video analytics. Video analytics technology is incredible. It allows software to analyze an image and classify thousands of attributes. It can determine what color clothing someone is wearing, approximate their age and gender, and even identify whether a vehicle in view is a car, bicycle, motorcycle, or truck. Some systems can even read license plates. The potential applications of video analytics are remarkable.

This hospital was exploring a specific use case for video analytics: facial recognition. Every time a person entered the facility, the camera system would scan their face, similar to a fingerprint, and classify them. The committee's task was to determine whether this facial recognition technology should be used to identify known threats. For example, if a caregiver or employee at the hospital had been involved in a domestic violence case that posed a potential danger to them at work, the system could recognize the individual responsible and alert hospital security when they entered the building. This would allow staff to respond quickly, preventing harm to the caregiver.

It's a powerful tool, right?

(This technology can also be used in schools to detect known sex offenders or alert staff to the possibility of a sex offender entering the premises.)

It was entirely appropriate for the hospital to explore this technology, but the real power of this story wasn't in the technology itself—it was in the committee's process of evaluating it.

Without a committee, a security professional might say, "Yes, this tool helps me do my job. It helps me protect caregivers," and proceed with implementation right away. However, when you involve compliance and legal teams, you see a more balanced perspective.

Questions emerge:

Q. *Is this legal?*
Q. *Is this what we want to be known for?*
Q. *Does this align with our organization's values?*

Sitting in on these discussions was fascinating. It revealed the power of a committee where diverse perspectives were shared openly. Ultimately, the organization decided to pause the implementation and revisit it in a year. They wanted to wait until other organizations using the technology had established case precedents. By observing how others fared—through legal challenges or operational success—they could make a more informed decision, avoiding the risks of being early adopters.

As you navigate your role as a security professional, consider the value of including other perspectives. These voices can complement your skills, address your blind spots, and ensure well-rounded decisions that align with your organization's goals and values.

CREATE A FOUNDATION OF COLLABORATION

While committees are not perfect, the challenges they present are part of what makes them valuable. Yes, they move more slowly, but that deliberate pace allows for thoughtful discussion and better-informed decisions. In high-stakes, regulated industries, rushing ahead without consensus can lead to overlooked risks and costly mistakes. Embracing the slower process ensures a stronger, more sustainable outcome.

It's also natural for some individuals to feel a pull toward quick action. There's a temptation to bypass the committee process and just get things done. However, for complex security initiatives, taking the time to gather input from multiple departments is critical. These perspectives help catch blind spots, ensure compliance, and provide the balance necessary to manage both operational needs and regulatory requirements.

Of course, without executive support, even the best committee can struggle to gain momentum. A strong executive champion aligns the organization behind the effort, making it easier to secure resources and cooperation from other departments. Without that backing, committees risk stalling before they've made meaningful progress. Identifying and engaging an executive sponsor from the start is essential to overcoming this challenge.

Ultimately, committees require patience and persistence, but the payoff is well worth it. They create a foundation of collaboration that builds trust across departments, ensures compliance, and drives long-term success. When each department's voice is heard and decisions are made thoughtfully, the result isn't just a functional security system—it's a solution that serves the organization's needs and values, with buy-in from every corner of the team.

Now that we've assembled our cross-functional security team, it's time to put their collective expertise to work. In the next chapter, we explore the first tool in the SAFE toolset to implement a modern security system, starting with S for Survey.

CHAPTER 4:
S IS FOR *SURVEY*

had the opportunity to work with a client tasked with managing the security system for a medical services business going through a rapid growth and expansion plan. Their plan included buying other medical facilities and opening new locations in a short period of time. The security executive was responsible for ensuring the implementation was consistent, met the organization's needs, and covered its risks.

We began by forming a committee to survey the needs of the business. Then, we worked together to bring in all those individuals with unique perspectives to discuss existing risks. These weren't the fastest meetings, but they uncovered a lot of concerns and needs that the security system needed to address.

Through the survey, we gained a clear view of the landscape. We saw where things could go wrong and what risks we needed to mitigate. Once we had that survey, it became apparent that we could set guidelines and create a security plan to ensure successful openings of new locations and integration of acquired businesses into their standards.

In this chapter, we're going to explore the critical first steps in modernizing your security system approach. These steps include surveying risks to pinpoint risk and uncover hidden vulnerabilities, setting guidelines and desired outcomes for your security system, and establishing comprehensive standards that will guide your security strategy moving forward.

STANDARD SECURITY SYSTEM SURVEY

A standard framework for surveying a security system is typically defined in four parts. A security system should:

1. Deter or discourage people from taking any bad action against your location.
2. Detect and identify any unauthorized access.
3. Delay people from gaining unauthorized access.
4. Notify and respond to incidents quickly and effectively.

Using this layered approach (and a committee to evaluate your security needs), you can begin surveying the risks that exist within your business. When we work with clients, we usually start by looking at the physical building, asking questions about what's being protected, where assets are located, and how someone could gain access to those locations. We typically ask questions like:

1. What are your most valuable assets (e.g., data servers, cash rooms, product inventory)?
2. Where are these assets located within the building?
3. What current security measures (e.g., locks, cameras, access control) protect these areas?
4. Who has authorized access to these areas?
5. What are the current entry and exit points for the building?
6. Are there any less obvious access points (e.g., loading docks, emergency exits)?

7. What is your current visitor management process?
8. Have you experienced any security incidents in the past?
9. What are your hours of operation? What periods are your buildings unoccupied?
10. Are there any specific regulatory requirements your security system must meet?

This list provides a starting point for a thorough security assessment, helping identify vulnerabilities and prioritize protection measures.

We can start defining strategies to deter, detect, delay, and respond to incidents by taking a layered approach to your facility, starting at the target or center.

Once we've defined your layered approach to protecting the targets in your building, we typically begin a risk evaluation. This evaluation lists all the possible threats and each way somebody could gain access, ranking the severity.

HOW TO SURVEY AND SET GUIDELINES

So, how do you begin surveying your facility and putting together guidelines for a consistent, modern security system? Step by step.

STEP 1: SURVEY YOUR BUSINESS

Knowing what to secure first means surveying your business. For surveying, we have resources (as do industry associations) for conducting security assessments that typically use the layered approach we've discussed. These comprehensive survey tools are quite extensive - often running 20-30 pages or more - which makes them too large to include in full within this book. However, I suggest you obtain one of these detailed survey tools for identifying your risks and their magnitude. They provide a structured, thorough approach to assessing your security needs.

STEP 2: DEFINE GUIDELINES

After the risk assessment survey, you need to define guidelines for your security system. Most new installations in the security industry use Division 27 and Division 28 of the engineering and design standards to define what a security system includes. (Division 27 governs building communication systems and Division 28 governs electrical safety and security systems.) Some businesses we work with go so far as to develop their own standard Division 27 and Division 28 that outlines what manufacturers they want to use and how systems should be installed. These standards are what would be handed to an installer, defining where devices need to be placed, what type of card reader to use, what manufacturers are authorized, and what type of encryption is on the card reader.

Other businesses take a more general approach and write a requirements guideline document. This is less technical and detailed than engineering specifications, but provides general objectives that your security system should achieve.

An example is video surveillance recording standards. When setting up your video system, you have choices: You can record continuously or only when there's motion. You can define how long recordings are stored: 7 days, 14 days, 30 days, 90 days, etc.

Here are some other questions to ask:

1. What software will you use to manage video recording?
2. What redundancy do you require in your recording system?
3. Do you need the ability to lose power on one supply but keep the system working?
4. Do you need redundant operating system drives or a RAID storage configuration?

All these tools help ensure video isn't lost, but they come at a cost. A user requirements document can provide general outlines on the standards you want to set, without going into detailed engineering specifications.

STEP 3: START WHERE THE NEED IS THE HIGHEST

Next, start where the need is highest. If your most pressing issue is protecting a specific asset, start by defining a security standard, assessing the risk of that site or asset, and setting standards for how to do that.

STEP 4: SET REASONABLE GOALS

The last how-to tip is: don't try to take on too much too fast or set overly ambitious goals. Set *reasonable* goals for what you want to accomplish and in what timeframe, then begin surveying and setting your guidelines.

An example might be setting a standard for your intrusion alarm system. A realistic expectation could be that within three months, you've defined how much intrusion alarm you want in your buildings, what types of buildings should have intrusion alarms, to what extent you want to install intrusion alarms within buildings, and what manufacturers and installation guides you require. That's a realistic expectation for what you and your committee can tackle in three months.

CASE STUDY: AN INTERNATIONAL ACQUISITION SURVEY

An example of successful use of this survey and guidelines approach was with an international chemicals company doing mergers and acquisitions.

They purchased another large business with over 100 facilities to integrate. Through this acquisition, they inherited the strengths and weaknesses of the acquired business's security standards and needed to survey 100 facilities.

They set up a simple, easy-to-use questionnaire to distribute to the facilities. The questionnaire asked for a point of contact for security system management at each site and posed questions like:

- Do you have an intrusion system?
- A card access system?
- A video surveillance system?
- What manufacturer is it?
- Is it used throughout the facility or only in specific areas?

This survey was very successful, identifying that their international regions were using an unauthorized video surveillance system banned by the US government.

Remember the multinational chemicals company we discussed in Chapter 5? Their situation provides another excellent example of the benefits of clear security guidelines. This company sells to the US government and must comply with restrictions against using any banned Chinese video surveillance systems in their network.

During their acquisition process, their survey revealed that some international facilities were using unauthorized surveillance systems. Thanks to their well-defined security standards, they quickly identified this compliance issue. The company was able to swiftly instruct the newly acquired business to disconnect these systems, ensuring immediate compliance with government standards.

This case further illustrates how having clear, comprehensive security guidelines can help navigate complex situations, especially during mergers and acquisitions. It allowed the company to act decisively, avoiding potential legal and contractual issues with their government clients.

CASE STUDY: SECURING RADIOACTIVE MATERIAL

When dealing with high-risk assets like radioactive materials, it's crucial to bring the right people into the conversation from the start. This was especially true for a leading nuclear research facility we worked with. The company conducts critical experiments and houses significant amounts of radioactive material, making security a top priority.

The stakes were incredibly high. Any security breach could lead to theft of radioactive materials, potentially endangering public safety or even enabling the creation of dirty bombs. Additionally, unauthorized access to research data could compromise national security and the company's competitive edge.

We started by assembling a cross-functional team. Each group brought unique expertise:

- IT flagged cybersecurity risks, crucial for protecting sensitive research data
- Facilities ensured continuous power, vital for maintaining containment systems
- HR managed biometric data collection and privacy, balancing security with employee rights

- Nuclear safety experts advised on regulatory compliance and material handling protocols

Our survey process was meticulous. We conducted physical walk-throughs, reviewed existing security measures, and analyzed potential vulnerabilities. We examined everything from perimeter fencing to internal access controls, considering both physical and cyber threats.

Developing standards required thoughtful collaboration. We had in-depth discussions about:

- Advanced biometric systems for access control
- Radiation detection equipment integration
- Tamper-evident seals for sensitive areas
- Redundant power systems for critical security components
- Secure communication protocols for emergency situations

The result wasn't just a security plan for one site. It was a comprehensive deployment guide for all the company's facilities handling radioactive materials. This ensured consistency and reliability across the board, crucial for maintaining the highest security standards.

This project showcases the power of a well-rounded team. By involving the right experts and setting clear standards, we created a robust security framework that not only prevents threats but also adapts to evolving challenges. No single department could have achieved this alone, but together, they built a system that protects the business, its assets, and potentially, national security.

PINPOINT RISKS AND ESTABLISH PRIORITIES

Performing thorough risk assessments and building a solid security plan are essential for identifying vulnerabilities, protecting your organization, and ensuring regulatory compliance. A well-executed survey is more than just a checklist—it's the foundation for a security strategy that can grow with your business. Whether you're integrating new acquisitions or securing high-risk assets like radioactive materials, success comes from careful planning, setting clear standards, and working with the right people.

Surveying your security landscape helps pinpoint risks and establish priorities, but the real power lies in setting guidelines that ensure consistency across locations. These standards—whether they're technical frameworks like Division 27 and 28 or broader user requirement documents—serve as a playbook for protecting your assets and managing risks effectively.

The example of the international acquisition highlights why having clear guidelines matters. When the survey revealed the use of a banned video surveillance system, the company was able to act fast—disconnecting it to stay compliant with US government standards. Without those clear rules in place, that issue might have gone unnoticed until it was too late.

The case study on radioactive materials shows how important it is to bring in multiple perspectives to build a system that works under pressure. No one department can handle every challenge alone, but together, they create a plan that's both strong and adaptable.

Now that we've surveyed the landscape, identified vulnerabilities, set priorities, and established standards, it's time to put this plan into action. In the next chapter, "A for Apply," we'll dive into how to implement these insights and turn your security blueprint into a functioning defense system.

This is where all the planning pays off. It gives you the tools to stay ahead of evolving threats and build a security framework that protects and grows with your organization.

CHAPTER 5:
A IS FOR *APPLY*

etting your security system installation standards effectively installed is not an easy feat. One experience we had was serving as a consultant during a building upgrade. The customer had been through the process of setting up their standards and asked us to help them apply those standards at this building. First, we read the specifications that the architect or engineer had documented.

The client sent them to us and said, "Could you review these and ensure that they're in compliance with our standards?"

We found the wrong camera manufacturers were referenced as we went through it. They were *not* compliant with their standards, and we were able to get that addressed and corrected before they got too far down the installation process.

When we reached the point where the electrical contractor doing the installation was selected, they submitted their plan for the hardware they were going to install. They did a great job reading the specifications, which called for license plate reading cameras in the parking lot. However, they chose to install large 24-inch cameras on small parking lot poles. These cameras are typically used on freeways or at toll booths—way too big for the task at hand. To make matters worse, they were not only overkill but also incredibly expensive and high-performance, far beyond what was needed.

Fortunately, we were able to step in and save the customer almost $1,200 per camera by recommending a more appropriate option that also looked much better. This decision saved the client $12,000 across 10 cameras while ensuring the system was still fully functional and effective.

The last challenge this site faced was with the cabling installation for the parking lot poles. Upon reviewing the cabling, we discovered that the cable runs were over 700 feet. CAT 6 data lines can't transmit data over that distance without losing packets of information, which would cause the video to degrade or be lost entirely. We advised the client on the appropriate cabling to ensure the pole cameras would function properly over those distances.

I share this story to highlight challenges and roadblocks you may face throughout an installation process and the value of having experts support you through the installation.

In this chapter, we will discuss the importance of assembling a comprehensive installation team of security experts who can apply your modern security system.

KEY PLAYERS IN APPLYING YOUR
SECURITY SYSTEM

As you begin implementing your security system, I'd like to start by identifying the different individuals who will assist you throughout the process. The specific people or roles may vary, but this will give you a good idea of the types of functions and expertise you'll be working with while applying your new modern system.

SECURITY COMMITTEE

Your security committee is a resource throughout the implementation. As questions arise or issues come up while you apply your new system, they're an ongoing resource.

ARCHITECTS AND ENGINEERS

The second group you'll likely work with are architects and engineers. Constructing a new building or doing a major remodel needs experts drawing architectural plans that encompass HVAC, mechanical, plumbing, and electrical systems, flooring, ceiling grids. It's complex, and your business will leverage an architect and an engineer.

Your security system fits within some of those specialized systems that an architect or engineer will define or document in the building plan. (As mentioned previously, Division 27 governs building communication systems and Division 28 governs electrical safety and security systems.) An architect or engineer will ensure that divisions

27 and 28 accurately represent the security standards you want implemented, working together with the architect and engineer.

SECURITY CONSULTANTS

Another group you might work with are security consultants. Not all architecture and engineering firms have deep experts in security systems. Security consultants exist as a reference to ensure standards for building installations are defined, inspected, and properly implemented. A security consultant will know not only general information about a security system but also specific part numbers. They will review installation plans from the awarded contractor to ensure compliance with the installation standards.

SECURITY SYSTEMS INTEGRATORS AND ELECTRICAL INSTALLERS

The third group you'll probably interact with are the physical installers. Security systems are usually installed by one of two parties. One is a security systems integrator. Security systems integrators are usually capable of doing the end-to-end installation—cabling, device installation, and programming and configuration. This group is unique or powerful because they have deep expertise on security systems specifically.

The second party you might interact with on the physical installation is an electrical contractor. An electrical contractor often does most of the cabling within your building, including the data and security systems installations. Since they're already running other cables, there's value to them also doing the security system cabling and installation. They're a great resource—they have scale, speed, and can tackle the physical installation.

In large system installations, it's common for the security systems integrator to provide the parts (and the smarts) to install the system, and for the electrical contractor to provide the labor to install it.

SYSTEMS CONFIGURATION

The last group you are likely to interact with is systems configuration. Once a security system is installed, it needs to interact with your building and business systems. You have a choice on whether to let other companies access your network to add the camera or card access system to your active directory database or do that yourself. This specialized function of merging the security system into your building system is a unique role that you need to plan for to ensure a successful installation.

Applying your security system takes a village.

TIPS FOR SUCCESSFUL IMPLEMENTATION

Now that we've defined the parties you'll be interacting with while applying your security plan, I'd like to share some tips.

TIP 1: DEFINE YOUR ORGANIZATION'S RESPONSIBILITIES AND CAPABILITIES

You can bring that expertise in-house if you or someone on your team is a deep security expert and has the knowledge of manufacturer part numbers and installation standards. If you're confident in those skills, there's no need to outsource. However, if this is an area where your organization is lacking, you may need to bring in a security consultant, architect, or engineer to support you where you're weak. A quick capability and responsibility assessment is key, and I recommend doing this across the five responsibility areas we discussed in the Key Players section.

Ask yourself:

- Will your organization handle cabling and device installation in-house, or will you need an outside expert or installer?
- Do you have a security expert or consultant internally, or will you need to hire one?
- Will architects and engineers define your security standards and put them out to bid?

Finally, is there someone on your IT team who can handle the final setup for systems configuration, or do you need an external vendor for that?

Laying out these roles and responsibilities will help you understand where your organization is strong and where you may need to bring in additional expertise and support.

TIP 2: CONTINUE TO DEVELOP YOUR IMPLEMENTATION PLAYBOOK

We've talked about having your security standards or guidelines that can evolve into a well-defined Division 27 and Division 28 specification. This way, you don't have to rewrite them every time you do a building installation; they can be predefined and circulated to the next architect and engineer.

Continue to document and refine those standards, so you can repeat them in each facility you go to.

TIP 3: WORK WITH A SECURITY EXPERT AND SYSTEMS CONFIGURATOR

I cannot stress enough the importance of having a security expert and systems configurator on your team. Here's why:

A security expert knows all about current threats and rules you need to follow. They can spot where you're at risk and design a system to protect you. This helps avoid making expensive mistakes when choosing and setting up your security.

A systems configurator makes sure all parts of your security system work well together. They set up cameras, alarms, and door locks so they all communicate with each other. This means you get the most out of the money you spend on security.

Together, these people help:

- Find and fix risks
- Follow all the rules
- Design the best system for you
- Fix problems and make the system better over time
- Teach your staff how to use the system

Without them, you can end up with a system that doesn't work well or leaves you open to attacks. In today's world, where security threats are always changing, you need these skills to keep your business and people safe.

CASE STUDY: A MULTINATIONAL CHEMICALS COMPANY'S APPROACH

A company we work with has a deep expert in physical security systems. This individual is a board member of a security systems association. They stay current by reading trade publications, meeting with manufacturers, and knowing the company's products inside and out. As a result, the company's security committee has developed specific Division 27 and Division 28 guidelines.

The company has a team of six in-house IT systems engineers. These engineers handle the final configuration of newly installed security systems, integrating them into the company's network. They are also responsible for maintaining the system's ongoing operations.

In this case, the company has chosen to cover its bases with the deep security expert consultant *and* an in-house team for systems configuration and maintenance. They've invested in these resources full-time to ensure consistent application and upkeep.

As you assess your organization and the strategic plans for physical security, you need to determine if you want to do this internally by hiring the full team needed to support security or external experts.

MANAGING SYSTEM CREEP

As you're applying and implementing your security standards, the last thing to be mindful of is that creep will happen. If you are a large organization, you have multiple stakeholders managing buildings and interacting with the systems and multiple architects and engineers involved in the buildings. In spite of your best efforts, things can start to creep into your system.

For example, an operations manager may need to have a video surveillance system that's viewing a specific process or system inside their facility. Instead of consulting the corporate standards, they go out and buy their own system. Whether it's from a big box electronics company or by asking another security systems integrator to bring in a system and install it for them, it's an example of systems creeping into your building that may not comply with your standards.

Here are a couple of suggestions to monitor and maintain your system's compliance:

- **MAKE SURE YOU KNOW THE SECURITY CONTACT** at each facility that you oversee. Build good relationships with these people. They are your eyes and ears in the facility throughout the year.

- **MAKE OCCASIONAL IN-PERSON VISITS** to those experts and the organization, so you can be mindful of what's happening within the facility. It builds a relationship of trust, so if they see something happening, they're more comfortable or confident to call you.

- **MAKE SURE TO CIRCULATE YOUR SECURITY STANDARDS**, as they are meant to guide the organization. Each onsite facility representative overseeing the security system should know your guidelines and have a copy of them.

Periodically, we recommend sending out questionnaires to the facilities and asking the security contact to explicitly answer questions about what might be considered creep within your system. It also shows if they're using and are trained on the system. An example of this would be asking the building security manager if they are setting their intrusion alarm every night. You want to know if they have a system that's being used, the last time people were trained on it, and the last time passwords were updated.

Managing system creep is a challenge that even the best-laid security plans can't completely avoid. In large organizations with many moving parts, it's easy for unintended variations to slip in—like an operations manager bypassing corporate standards to install their own camera system.

Build solid relationships with your on-site security contacts and make regular visits to see what's happening on the ground. Share your security guidelines widely and ensure everyone has a clear understanding of what's expected. Regular check-ins, like questionnaires, can help you catch issues early—before they turn into bigger problems.

You can maintain a security system that stays consistent, reliable, and true to your standards by monitoring potential deviations and addressing them as they arise.

ACTIVATE YOUR SECRET WEAPON

Your secret weapon to ensure your security plan is correctly implemented is an adept team of diverse security professionals. As we saw with the building upgrade example, having the right people in place can make all the difference, whether it's catching mistakes early or ensuring the correct equipment is installed. By bringing in the right mix of expertise, from security consultants to engineers and systems integrators, you can be confident that your security system will be built to last, serving your needs for years to come.

But remember, you can't overlook managing system creep. Over time, things can start slipping into your system that don't meet your standards, whether it's someone installing a different camera system or bypassing your guidelines. By staying connected with your on-site contacts, conducting regular reviews, and making sure everyone has the latest standards, you can keep your system on track and avoid unwanted surprises down the road.

Now that your security plan is in action and standards are being applied across the organization, it's time to take your defenses to the next level. In Chapter 6, we will explore how to strengthen and reinforce your security measures, ensuring your newly implemented system can withstand and adapt to evolving threats.

CHAPTER 6:
F IS FOR *FORTIFY*

Our company has supported a jail as part of our services for several years.

When I first became involved with our business, I noticed that every morning a technician accessed the camera system at the jail to view every image, one at a time, to see if the camera was working and recording.

The jail prioritized the reliability of the cameras. We had clear service level expectations and needed them to perform well. At the time, the most reliable approach was to have someone manually log in, identify cameras that weren't working, and create an action plan to address the issues they found.

But watching this made me think, *"There's got to be a better way."*

Technology systems excel at tasks like this. A machine can easily detect if something is transmitting data, determine if it's online, and analyze the image being transmitted to decide if it's a good or bad image. That realization made me wonder: how can we maintain this client's system more efficiently and at a larger scale?

After speaking with other businesses, I realized we weren't alone in our approach. For example, a university had an IT resource log into their camera system daily, scanning hundreds of camera views to ensure they were recording properly. A multinational company was doing the same—someone manually logged in, documented issues, and addressed them. Depending on the size of the system, this manual process was taking businesses anywhere from one to four hours per day.

Plus, there were delays since these checks happened once a day, at a specific time. A machine doesn't need a lunch break or a good night's sleep; it can continuously monitor and watch over your security system.

In the next section, we'll build on this example, discussing the tools we've implemented to maintain security systems more effectively for our clients. This chapter will guide you through essential proactive security measures that keep you safe in an ever-changing threat landscape.

TRADITIONAL VS. MODERN SECURITY SYSTEM MAINTENANCE

I'd like to begin by defining how the traditional security system is maintained.

We've talked about how an electrical contractor or security systems integrator will install a system. But then what happens to it?

- How is it monitored for performance?
- How is it updated?
- How is it repaired when issues are detected?

In the worst-case scenario—though not entirely uncommon in the industry—the system gets installed, there's an assumption it will keep running, and no proactive steps are taken to maintain it. In the traditional security approach, there's often no obligation for an outside company or security expert to support the system once it's in place. Sometimes, companies handle this themselves, dedicating resources to monitor and support their security systems.

But the reality is that after a system is installed, it immediately begins to depreciate, decline, and degrade—much like a car that starts losing value the moment you drive it off the lot. Just as a car experiences wear and tear, a security system does too. Regular oil changes keep a car running smoothly despite some decline, a similar approach is needed for security systems. In the traditional security model, action is often only taken when the system reaches a critical issue, and only then is a team mobilized to fix it.

COMPARISON OF SYSTEM PERFORMANCE OVER TIME:
Traditional vs. Modern Security Systems

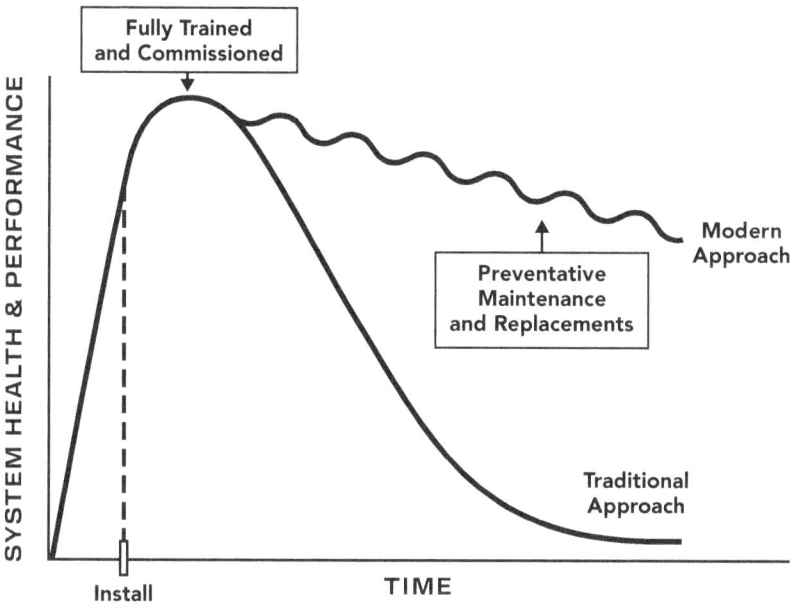

This graph illustrates the performance of traditional security systems compared to modern, proactively maintained systems over time.

When I think about how most people treat their security systems, I'm reminded of the way folks used to treat old house locks. You stick one on the door and assume it'll work forever. No maintenance, no updates—just trust it to keep doing its job. But security systems, especially today's more complex ones, don't work that way. They need upkeep, just like a car needs an oil change. That's what this graphic shows. On the left side of the chart, the Y-axis represents how well your system is functioning—how healthy and reliable it is. Along the bottom, the X-axis tracks time. When a traditional security system is first installed, everything runs great— top-notch performance, everything working as it should. But then, over time, without maintenance, performance steadily declines. Cameras fail, recordings get lost, systems lag—and most people don't notice until something critical breaks. That's the bottom line.

Compare that to a modern, proactively maintained system. It starts strong too, but as normal wear and tear sets in, you address issues early. You replace devices before they die, keep the software updated, and check the system regularly. Performance might dip here and there, but you catch problems before they snowball. That proactive approach is the difference between "letting that baby die" and actually keeping people and property protected. It's not just about installing a system—it's about keeping it healthy, day in and day out.

The goal of the modern approach is to keep your security system operating as close to optimal as possible. Let's dive into the actions you can take to maintain that high level of performance.

ACTION 1: CONTINUOUS HEALTH MONITORING

First, let's talk about continuous health monitoring. This involves using software to watch your security system for specific data and alert you when an issue arises. In the best-case scenario, if your system is advanced enough, it can take steps to resolve issues automatically, without needing human intervention.

A well-known, though controversial, example of what can happen without continuous monitoring is the Jeffrey Epstein case. If you're unfamiliar with the story, at the time of Epstein's death—a high-profile inmate in a jail cell—the camera outside his cell stopped working. Other cameras in the area also appeared to have corrupted footage and were deemed unusable. According to investigation reports, it remains unclear how long these devices had been offline before the incident. They could have been down for days. The hard drive where the footage was stored have been corrupted for a while. In a modern security system, all this data would be logged, recorded, and made available for an investigation.

The example has fueled countless conspiracy theories.

People ask: Was the system intentionally disabled at that moment, or was it the unfortunate state of an outdated system?

I can't provide a technical answer for what happened, but I can tell you that, in my experience visiting many prisons, it's not uncommon for them to use outdated video surveillance systems. Seven out of ten facilities I've visited lack fully functional systems, and it's not rare for some cameras to be down. When investigations rely on video footage, that footage is often unavailable. So, it's within the realm of possibility that this jail's security system was down without any conspiracy involved.

Continuous health monitoring could have made a world of difference in understanding what really happened.

ACTION 2: PASSWORD UPDATES

Another key part of proactive system management is keeping device passwords updated. If an installer sets up a system and uses their password—and it stays the same for 15 years—that password ends up being known by a lot of people. The best practice is to update those passwords regularly to better protect your security systems.

For example, take a recent data breach involving Hikvision cameras. Hackers reportedly accessed over 80,000 cameras. In the process, they gathered all the usernames and passwords for those devices. More than 50% of the passwords were either the default ones provided by the manufacturer, easily found online, or basic passwords like "companyname1234." It's a huge vulnerability. Using complex, randomly generated passwords—through tools like LastPass, 1Password, or whatever system your company uses—can make a big difference. A unique password that's not shared widely within your organization keeps your systems much more secure.

ACTION 3: FIRMWARE MANAGEMENT

In 2017, hackers broke into Equifax (a major credit reporting company) and allegedly stole personal data from 147 million people. The fallout was a $700 million settlement and a reputation for having one of the biggest data breaches in history.

How did it happen? They left a known security hole open for months, even though a patch was readily available.

Keeping up with firmware updates is an essential tool for proactive system management. These updates help minimize cybersecurity risks, enhance device performance, and introduce the latest technological features. From my experience conducting physical security assessments, it's rare to find organizations updating their firmware annually—most do it every few years or not at all. The best practice is to regularly monitor and update the firmware on your devices.

Why doesn't this happen more often? Sometimes, it's because no one is specifically responsible for applying updates. Other times, it's simply because there's no policy in place—if it's not required, it doesn't get done. The solution? Either hire an outsourced expert or assign someone in-house to stay on top of updates. You need a proactive schedule for applying firmware updates. The updates come from the manufacturer, and once applied, it's important to double-check that the device is still working properly.

Don't let your company be the next headline. Stay updated; stay secure.

ACTION 4: SECURITY SOFTWARE UPDATES

Staying on top of software version updates for your security system is essential. These updates ensure your video surveillance system remains compatible with its operating environment—whether it's Windows, Linux, or another system. Major releases often include critical patches and new features that enhance the system's performance. Just like with firmware updates, it's crucial to align these software updates with any changes in the operating system to keep everything running smoothly.

So, why aren't these updates more common? Often, it's because no one is specifically tasked with managing them, and there's no clear plan or schedule for how often they should occur. To address this, we recommend scheduling these updates at least once a year and staying vigilant about applying security patches or hot fixes as they become available.

A modern security system approach ensures that the operating system, application updates, and firmware versions are updated regularly. We all consistently update our mobile phones, laptops, and software applications. Your physical security system should be treated with the same ongoing maintenance.

ACTION 5: HARDWARE INSPECTIONS AND PREVENTIVE MAINTENANCE

Another effective tool for ensuring your system's operation is doing hardware inspections and preventive maintenance. Think about the door locking system that's securing your building. It's used every day. That wear and tear can make it loose. It can make the door not close properly and should be inspected, adjusted, and made sure that it's working reliably. Put this in your preventive maintenance inspection plan for your internal building maintenance team or with your trusted partner.

Similar to that are battery replacements. If you lose power in your building, the door access system and the intrusion alarm system rely on batteries to keep the doors secure. These lead acid or lithium ion batteries wear out over time, so you should have a preventive maintenance plan to inspect and replace those batteries as needed.

Another best practice is camera lens cleaning. If you're in a manufacturing setting or in an outdoor environment, your lenses get dirty. The image quality goes down when the dome or the lens gets covered in oil, grease, or dirt. Doing lens cleaning ensures your camera is working appropriately.

For a checklist of these activities and a recommended interval to do them, see Appendix I.

TYPICAL SERVICE INTERVALS

This table outlines a standard maintenance schedule for modern security systems, broken down into remote service and onsite service tasks. Each activity is assigned a typical service interval to ensure optimal system performance and reliability.

This table outlines a standard maintenance schedule for modern security systems, broken into two main categories: remote service and onsite service. Under remote service, monthly operating system updates ensure that all patches and versions are current. Annually, security applications should be updated to the latest version, and all critical passwords should be reset. Additionally, user access audits should be conducted quarterly to remove inactive accounts, passcodes, or badges that no longer need access.

Several key components on the onsite side require annual attention.

Alarm systems should be tested to verify all alarm points and their connection to the monitoring center. Door hardware—including locks, closers, sensors, and card readers—needs to be tested for full functionality. Surveillance cameras should be cleaned and their aim adjusted to maintain optimal coverage. Lastly, battery health should be evaluated annually to ensure the backup power supply remains reliable in the event of an outage.

Let's break it down.

TYPICAL SERVICE INTERVALS FOR SECURITY MAINTENANCE

	ACTIVITY	DESCRIPTION	FREQUENCY
REMOTE SERVICE	OS UPDATES	APPLY ALL PATCHES AND VERSION UPDATES	MONTHLY
	APPLICATION UPDATES	UPDATE TO THE LATEST VERSION OF THE SECURITY APPLICATION	ANNUALLY
	PASSWORD RESETS	RESET ALL CRITICAL PASSWORDS	ANNUALLY
	USER ACCESS AUDITS	DEACTIVATE INACTIVE USER ACCOUNTS, PASSCODES AND BADGES	QUARTERLY
ONSITE SERVICE	ALARM SYSTEM TESTS	TEST ALL ALARM POINTS AND MONITORING CENTER COMMUNICATION	ANNUALLY
	DOOR HARDWARE SERVICE	TEST ALL LOCKING HARDWARE, CLOSERS, DOOR SENSORS, CARD READERS	ANNUALLY
	CAMERA AIMING & CLEANING	CLEAN ANY LENS AND ADJUST THE LENS DIRECTION	ANNUALLY
	BATTERY TESTS	TEST VOLTAGE AND BATTERY HEALTH	ANNUALLY

REMOTE SERVICE TASKS

These are tasks that can be handled offsite, often through central-ized software platforms:

- **OPERATING SYSTEM (OS) UPDATES**: Apply the latest security patches and version updates monthly to ensure devices stay protected against emerging threats.

- **APPLICATION UPDATES**: Upgrade security software to the most current version annually, keeping functionality and compat-ibility up to date.

- **PASSWORD RESETS**: To reduce the risk of unauthorized access, critical system passwords should be reset annually.

- **USER ACCESS AUDITS**: Conducted quarterly to identify and deactivate inactive user accounts, expired badges, or unused access codes.

ONSITE SERVICE TASKS

These require a physical visit to the facility to ensure the mechani-cal and hardware components are fully functional:

- **ALARM SYSTEM TESTS**: Annual testing verifies that all alarm points are operational and properly communicate with the monitoring center.

- **DOOR HARDWARE SERVICE**: All locking mechanisms, sensors, and card readers should be checked yearly for proper function and wear.

- **CAMERA AIMING AND CLEANING**: A yearly lens cleaning and directional check helps maintain clear image quality and proper surveillance coverage.

- **BATTERY TESTS**: Annual voltage and health checks on backup batteries ensure system uptime during power outages.

Keeping up with this schedule helps reduce the risk of system failure, ensures compliance, and provides peace of mind that your security infrastructure is doing its job.

ACTION 6: STAY INFORMED OF NEW FEATURES

We've given guidelines and recommendations on how to maintain your security system, so that it isn't degrading over time.

Another way to fortify your system is to stay informed of new features being developed and released on your security system that may better protect your building. For example, the camera system you installed last year has added features and benefits you can take advantage of by doing a simple update.

I've always been impressed with one of our clients who has an in-house security expert who also serves on the board of advisors for an industry association. The expert spends time with all the key manufacturers, learning about the features and capabilities of the system he oversees. That way, he is always implementing continuous improvements and making sure his team is trained to get the most out of the system.

ACTION 7: INVOLVE SECURITY COMMITTEE STAKEHOLDERS

The last recommended action for fortifying your security system is to keep your security committee stakeholders involved. As your organization grows, your security system needs to keep up. One common thing we see is that changes to the IT network —like a new firewall or an update to the threat detection system—messes with access to the security system. Sometimes, no one notices until there's a problem.

That's why it's on you, as the security executive, to stay in the loop with your committee stakeholders. Make sure you know what updates and changes they're making that could impact your setup. If HR is changing their badge policy or making a shift in how they handle biometrics, you need to know about it. That way, you can update your plan and make sure everything stays compliant.

FROM REACTIVE TO PROACTIVE

Remember the jail at the beginning of the chapter? We used to log in every morning to ensure the cameras were recording. Since then, we've implemented a continuous health monitoring system. Now, every 60 seconds, it checks the health of every camera, card access panel, and server component, collecting millions of data points on the system's health. Instead of having someone manually log in and inspect each image, the system automatically detects when an image is bad or a camera goes offline.

We've developed automation to reset a camera (in some cases) to see if a restart will bring it back online. If that doesn't work, the system automatically generates a service ticket for us to go on-site and fix issues like a failed hard drive or other detected problems.

This setup makes better use of a skilled IT resource, who no longer has to waste time logging in and checking images. It also provides a much better service for the customer. They have 24/7, 365-day health monitoring that takes proactive steps to fix problems. If they need to know how long a device has been offline, we have all that data logged and can provide a full report on the system's performance history.

Transitioning from reactive to proactive security measures can make a significant difference in today's rapidly changing threat landscape. As we saw with the shift from manual checks in the jail system to continuous health monitoring, adopting advanced technologies allows us to detect issues faster and take action before problems escalate. This shift improves efficiency, reduces vulnerabilities, and ensures a more reliable, 24/7 security presence. With continuous monitoring, automated responses, and well-documented system logs, you gain the upper hand in any situation.

The key to a successful transition lies in maintaining active engagement from key stakeholders. Regular meetings with your security committee keep everyone aligned on changes, whether it's a new firewall or an updated HR policy on badge access. This ensures that your security system adapts alongside your organization's needs. By incorporating routine updates, ongoing training, and industry insights, you prevent your system from becoming outdated and keep your security measures sharp.

Now that we've fortified our security systems with these robust measures, it's time to look ahead. In the next chapter, "E is for Evolve," we'll dive into strategies for keeping your security system up to date and adaptable, ensuring it remains effective and ready to meet whatever challenges the future holds.

CHAPTER 7:
E IS FOR *EVOLVE*

A few years ago, I had the chance to visit a stunning, state-of-the-art medical manufacturing building. The place was incredible—designed to bring customers in and showcase their cutting-edge medical equipment. It projected the power, presence, and expertise of the organization. Everything in the building was ultra-modern. As I walked down the hallway, I noticed the drinking fountain—one of those sleek ones you see in airports—but this one had both hot and cold water options. I thought, "Wow, they've really spared no expense here."

The phone systems were cloud-based, the lighting was all LED, and there were sound-dampening speakers throughout the building. I imagined they had the latest firewalls and top-notch cybersecurity in place too. As I continued down the hallway, right next to that fancy drinking fountain, something caught my attention.

Since I'm in the security space, I noticed that all the security cameras were about 20 years old. I later found out the card readers we were using to badge in were running on a Windows 7 operating system.

Despite all the investment in modernizing the building, they hadn't put the same effort into updating the *physical* security system. The cameras and card readers were outdated, leaving gaps in protecting their trade secrets, assets, and ultimately, the reputation of the business.

To help put the risk of obsolescence into perspective, it's crucial to consider the useful life of the security system you install. We're all familiar with the lifecycle of an iPhone—typically three to five years, depending on how eager you are for the latest features. But none of us expect to use an iPhone for 15 years. There's a limit to how long they'll connect to the network. If you've got an old 3G model, it just doesn't work anymore because the technology has moved on. It's the same with security systems—it's important to start by understanding the practical lifespan of your setup.

In this chapter, we'll delve into strategies that can make your security system resilient and adaptable to withstand obsolescence and changes in the security landscape.

HOW LONG SHOULD MY SECURITY SYSTEM COMPONENTS LAST?

TYPICAL LIFESPAN FOR SECURITY SYSTEM HARDWARE		
SYSTEM TYPE	DEVICE TYPE	AVERAGE USEFUL LIFE
VIDEO SURVEILLANCE SYSTEM	RECORDING SERVER	4-6 YEARS
	NETWORK SWITCH	5-7 YEARS
	IP CAMERA	7-10 YEARS
ACCESS CONTROL SYSTEM	CENTRAL SERVER/DATABASE	4-6 YEARS
	CONTROL PANEL	7-10 YEARS
	POWER SUPPLY	7-10 YEARS
	CARD READER	10-12 YEARS
INTRUSION ALARM SYSTEM	CONTROL PANEL	7-10 YEARS
	CELLULAR COMMUNICATOR	7-10 YEARS
	DETECTION DEVICES (E.G., MOTION, DOOR SENSOR, PANIC BUTTONS)	10-12 YEARS
INTERCOM SYSTEM	CENTRAL EXCHANGE SERVER	4-6 YEARS
	MASTER INTERCOM STATION	7-10 YEARS
	DOOR/REMOTE STATION	7-10 YEARS

The first step to resiliency is understanding the recommended useful life of the different components in your security system.

This graphic shows the typical lifespan of the major hardware components that make up modern physical security systems—like video surveillance, access control, and intercoms. I bucket them into three categories and break each one down by its key parts. For example, in a video system, there are the IP cameras themselves, which usually last 7 to 10 years, and then the recorder—the "brain" that stores all the footage, which typically lasts 5 to 7 years.

The same idea applies for access control. The control panels at the doors? Those can run for 10 to 12 years. But the main processing units? They need replacing sooner, usually in that 5 to 7-year window. Intercoms are a little more consistent, but they follow a similar pattern.

I like to use the iPhone as an example here. Even when a device still "works," not all parts age the same. Just like people try to hang onto their phones for too long, I've seen organizations push servers and recorders well past their intended lifespan. Then, they're surprised when the system fails right when they need it most. This chart is about helping people plan ahead—so they're not caught off guard by aging infrastructure that quietly stops doing its job.

These aren't hard and fast rules. If your equipment is in a particularly harsh environment, it may degrade faster. And if your policies call for more frequent updates, the timeline may shift.

This is the expected lifespan for most security equipment in today's market.

The big question is, when a business installs a security system, are they planning from the start to replace it in seven or ten years, depending on the equipment's life expectancy? In my experience, the answer is usually no. It's rarely built into the plan, but it should be. Documenting and planning for these replacements in advance ensures that your system stays up-to-date and performs securely, as we've discussed throughout this book.

I've seen companies neglect this planning too often. They let the system run until it simply stops working. Suddenly, they're facing an emergency purchase for a system upgrade—often at a higher cost, because it's done in a rush. This kind of scramble can throw the finance and operations teams into chaos, especially when there's no budget set aside for the expense. Unplanned and unbudgeted replacements are a real challenge, and it's one of your key responsibilities as a security executive to keep the organization prepared, knowing when replacements will be needed and providing an estimate of the costs involved.

KEEP UP WITH INDUSTRY TRENDS

To help you in that replacement lifecycle, either you or your outside security consultant need to stay abreast of the trends and advancements within the industry. A few tips for doing so are as follows:

TIP 1: ATTEND INDUSTRY TRADE SHOWS

To stay ahead in the ever-evolving security industry, it's crucial to engage with the latest trends and technology. A great way to do this is by attending industry trade shows. Make sure your security consultant is participating in these events to stay informed on new developments. One of the biggest gatherings in the U.S. is the ISC West Show, held each April at the Sands Casino in Las Vegas. This event is a hub for discovering new products and networking with industry leaders.

TIP 2: GET CERTIFIED

Beyond attending trade shows, you can earn certifications from recognized security associations like ASIS or the Security Industry Association (SIA), which are valuable. These credentials help ensure you're up to date with industry best practices and technological advances, reinforcing your expertise in the field.

TIP 3: SUBSCRIBE TO AND READ INDUSTRY PUBLICATIONS

Another key step is staying informed through industry publications. Subscribing to specialized magazines focused on security design and technology can keep you abreast of emerging trends and insights. Resources like these offer a wealth of knowledge that can help guide your decisions.

TIP 4: MEET WITH YOUR VENDERS

Additionally, regular meetings with your vendors are vital. These discussions provide opportunities to learn about the latest advancements in their products. For instance, connecting with your camera manufacturer can reveal new models or features, while conversations with your access control vendor can introduce you to the newest technologies they're implementing. These meetings ensure you leverage the most up-to-date tools for your security needs.

Staying updated through trade shows, certifications, industry publications, and vendor meetings ensures you're equipped with the latest security solutions. These steps help you keep your system effective, resilient, and ready to adapt to new challenges.

TIP 5: MEET QUARTERLY

To help our clients evolve, we stick to a quarterly meeting cadence. These meetings cover a few important things.

First, we talk about which systems are going to reach the end of their useful life in the next two years. This gives them plenty of lead time to budget for those replacements.

We also go over preventive maintenance, how the system is performing, and what updates or inspections have been done, thanks to our continuous health monitoring software.

Plus, we dive into industry trends—where things are headed, what new technologies they might want to think about. Then, as they see fit, they can make adjustments to their security plan based on those insights.

HOW THE SAFE FRAMEWORK SUPPORTS EVERY TYPE OF SECURITY EXECUTIVE

The SAFE Framework is designed to address the diverse challenges faced by security executives, adapting to their unique strengths while filling in critical gaps. Here's how the framework helps each type of professional tackle their specific risks, with real-world examples for context:

1. IT SECURITY EXECUTIVES

- **SURVEY**: Encourages them to assess physical vulnerabilities, like unmonitored doors or poor camera placement, beyond their usual focus on networks and cyber risks.

- **APPLY**: Provides a structured way to integrate physical security systems with IT infrastructure, ensuring a balanced approach.

- **FORTIFY**: Leverages their technical expertise by integrating IT monitoring tools with physical security maintenance, creating a unified system.

- **EVOLVE**: Ensures their systems—both physical and digital—stay current with evolving threats and technology.

An IT executive can automate checks for cameras, door systems, and badge readers using health monitoring software (discussed in Chapter 6). This complements their technical strengths while covering physical security and preventing blind spots like malfunctioning surveillance.

2. LAW ENFORCEMENT VETERANS

- **SURVEY**: Expands their focus by adding technical system assessments to their expertise in identifying physical threats and vulnerabilities.
- **APPLY**: Provides a framework for implementing modern systems like integrated access control and surveillance.
- **FORTIFY**: Moves beyond guard patrols to include structured maintenance of security technologies.
- **EVOLVE**: Helps them stay up-to-date with advancements in security technology, bridging gaps in their technical knowledge.

A jail case study demonstrates how a law enforcement veteran transitioned from manual camera checks to automated monitoring systems. The framework bridged their tactical background with advanced technology, ensuring better coverage and reliability.

3. CORPORATE SECURITY EXECUTIVE

- **SURVEY**: Encourages them to go beyond high-level assessments to uncover granular security gaps, ensuring both physical and digital vulnerabilities are addressed.
- **APPLY**: Helps them implement security systems that align with broader business objectives and satisfy operational needs.
- **FORTIFY**: Provides tools to measure and maintain security effectiveness, ensuring systems consistently deliver ROI and meet corporate expectations.

- **EVOLVE**: Guides them in keeping security programs current with emerging threats and technological advancements, maintaining alignment with the company's strategic goals.

A corporate office case study illustrates how a corporate security executive used the SAFE framework to justify a budget for upgrading outdated access control and surveillance systems. By incorporating metrics that demonstrated reduced downtime, improved employee safety, and measurable cost savings, they secured executive buy-in while maintaining alignment with business objectives.

4. FACILITIES AND OPERATIONS LEADERS

- **SURVEY**: Broadens their perspective to evaluate security holistically, including software and access controls, not just building hardware.

- **APPLY**: Provides tools to connect security systems with overall building infrastructure, streamlining management.

- **FORTIFY**: Integrates security system maintenance into routine facility upkeep, ensuring longevity and reliability.

- **EVOLVE**: Aligns security upgrades with other planned building updates, ensuring both evolve in tandem.

A medical facility case study shows how SAFE helped a facilities leader avoid overemphasizing building amenities, like locks and doors, while neglecting critical security tech upgrades. This balance improved overall safety without disrupting operations.

5. RISK AND COMPLIANCE PROFESSIONALS

- **SURVEY**: Incorporates practical security checks, such as blind spots or guard coverage, into compliance assessments.
- **APPLY**: Connects policies and regulations to actionable, real-world security measures.
- **FORTIFY**: Builds ongoing compliance tasks into daily security operations, ensuring standards are met consistently.
- **EVOLVE**: Keeps compliance aligned with changing regulations, avoiding obsolescence in policies or systems.

A case study of a chemical facility highlights how SAFE helped a compliance professional blend regulatory requirements with practical security improvements. By pairing policy updates with better access control and surveillance, they enhanced compliance and operational safety.

SAFE creates a robust, proactive security strategy that works for everyone by aligning these strengths and addressing gaps.

CASE STUDY: A NEW APPROACH TO SYSTEM MAINTENANCE

The jail we work with was tired of the constant cycle of equipment breaking down after reaching the end of its useful life, all without a solid budget plan in place, which made implementation a struggle. They decided to change their approach to implementing, maintaining, and updating their system. Instead of making a big one-time capital expenditure and then facing another down the road, they shifted their mindset to buying a service. This service guarantees them a working camera system, a functioning control system, a reliable intercom system, and the assurance that none of these components will reach the end of their useful life without a plan. The jail signed a 15-year agreement that includes at least two full system hardware replacements and continuous updates and support throughout that time.

This shift gave the jail administrators peace of mind. They no longer have to worry about whether their system is operational or struggle to persuade finance committees to invest in new equipment when the old one fails. Instead, they can focus on the care and safety of the inmates.

If you want to ensure that your system keeps evolving, consider not just buying the system outright as a one-time capital expense and then forgetting about future replacements. You can purchase it as an ongoing service or have a well-planned process for budgeting and planning for future upgrades.

It's an investment and a commitment to continuously keep your systems up to date. What's at stake if you don't is significant—there's real business profitability on the line, along with the risk of fines and damages. Keeping your system up to date helps demonstrate to your insurance company that you are appropriately managing your system. By staying on top of compliance standards and regulatory requirements through regular system updates, you're more likely to remain in line with or even exceed those standards, rather than scrambling to catch up when facing fines, penalties, or loss of business profits.

FUTURE-PROOF YOUR SYSTEM

Let's recap: your security system isn't a "set it and forget it" invest-ment. It has a finite useful life, and the key to keeping it effective is making sure it evolves over time. Planning ahead is crucial—whether that means knowing when your equipment needs replacing, adapting to the latest industry trends, or ensuring compliance with ever-changing regulations. This kind of forward-thinking approach keeps your system running smoothly, even as technology and security needs change around you.

Throughout this chapter, we've focused on the strategies to keep your security system fortified and future-proof. These steps help you stay ahead of the curve, from staying engaged with your security committee and conducting regular reviews to being proactive with updates and maintenance. It's about more than just keeping things working; it's about making sure your security measures can adapt and thrive in an evolving landscape.

But the journey doesn't stop at planning and maintaining. It's about using your security system's data to unlock new possibili-ties and drive business growth. As we move into our next chapter, "Data Alchemy—Transforming Security Metrics into Business Intelligence," we'll explore how to turn the data your security system collects into actionable insights. It's time to see how this wealth of information can do more than protect—it can shape your company's strategy and give you a competitive edge.

CHAPTER 8:

DATA ALCHEMY— TRANSFORMING SECURITY METRICS INTO BUSINESS INTELLIGENCE

was staggered to learn that there are 350 million IP security cameras deployed globally. Think about how much footage and data is being recorded around the world. To put that into perspective, it's one camera for each US citizen. And these cameras aren't sleeping. They're on *all the time*. All of that data has value. In this chapter, we're going to explore what some of the uses of that data are and how you can leverage it beyond just security.

I read a case study about a retailer using a security system originally set up to track theft and activity within their store. They started leveraging it for business intelligence. Most people don't realize how much effort retailers put into changing displays, adjusting product placement, and testing what sells better in different spots. Traditionally, they'd place a person with a clipboard to observe how customers moved around the store as product arrangements changed, and they'd track how retail transactions shifted based on these changes. In this case study, I was surprised to learn that the camera system meant for theft prevention was being repurposed for marketing research—tracking how long customers stopped at an end cap or paused to pick up a product.

Those cameras are now gathering valuable insights that drive sales and profits. It's exciting to see this shift—transforming data from being only about security into something that is a real asset and profit driver for the business.

This chapter will guide you on how to leverage your security data to transform it into insights and provide you with a competitive advantage.

TYPES OF DATA AVAILABLE FROM SECURITY SYSTEMS

Let's start with some of the types of data your security system can provide. First up are video images. These are incredibly useful for operations—tracking products, observing processes, and checking quality. As I mentioned earlier, cameras can monitor shopping patterns within a store, but they're also great for process improvements. They can highlight where inventory piles up or pinpoint wasted motions of employees or equipment on the floor. Another valuable use of video images is providing proof of shipping or loading products onto a truck. If there's a claim that something was lost or stolen, you can pull up the footage and say, "No, here it is—right there going onto the truck." Those video images hold real value.

Another type of data comes from card access activity. A badge or card is used to unlock a door, and it's also a record that tells you a lot about employee movement—when they arrived at work, when they left, and how much they moved around during the day. Some clients use the badging system for tracking time and attendance, so those logs are well documented.

We have another client using a device called a people counter. It tracks visitor counts, giving insights into event attendance and how effective advertising or promotions are in drawing people in. Originally designed for occupancy counting in security, it's now being used for retail analysis and attendance tracking.

ETHICAL CONSIDERATIONS IN DATA COLLECTION

Those are just a few examples of the kinds of data you can collect, but it's crucial to remember that all this information, like any security data, needs to be protected and kept secure. In many cases, you're dealing with personal information, so it's essential to follow your business policies on data retention, collection methods, and ensuring compliance with regulatory and corporate standards. Just because you can collect and analyze all this data doesn't mean it's always the right thing to do.

I'm reminded of an industry trade show where a CEO spoke about the rise of artificial intelligence and analytics in security. He referenced the Copenhagen Letter—a statement signed by several leading businesses and government bodies. It emphasizes that, as we push the boundaries of computing power and machine learning, having the capability to do something doesn't automatically justify its use. (A copy of the letter can be found in Appendix V.)

Keep this in mind as you guide the use of security data beyond just security needs. Use your security committee to help evaluate what's appropriate and ensure that you remain compliant. Thoughtfulness is key. Just because you can doesn't mean you should.

TIPS FOR IMPLEMENTING DATA VALUE

Let's shift gears to practical steps for turning security data into real value for your organization.

Start by reaching out across departments. Engage with teams like operations and find out what insights from the security system could help them. Be prepared with examples, such as where cameras are placed, what data is stored, and how industry trends are shaping the use of this data. These conversations can spark new ideas and lead to problem-solving opportunities you might not have anticipated.

Next, focus on quantifying the value of the data you collect. When you find ways to use security data that benefit other parts of the business, it helps shift the perception of the security system from a cost center to a valuable asset. Whether it's boosting efficiency, improving processes, or driving revenue through enhanced customer insights, documenting these gains helps the executive team see that investing in security has a broader impact beyond just keeping things safe.

THREE AREAS WHERE SECURITY DATA CAN ADD VALUE

Across industries, we've seen security data transformed into a valuable business asset in three areas:

- Improving customer experience
- Improving the quality of your products

- Increasing efficiency

Let's look at some examples.

IMPROVING THE CUSTOMER EXPERIENCE

An example of customer experience is using the camera system to test different retail spaces or product placement. This includes watching traffic patterns on the layout and flow of your floor plan, changing the lights in different areas to see where people dwell longer when the lighting is different, and changing the merchandise that's placed in specific locations. Mapping that video analytics with retail sales will be insightful as you grow revenue.

If this is a new concept, go to your camera manufacturer or video management software vendor and ask them what retail analytics developments are available.

IMPROVING THE QUALITY OF YOUR PRODUCTS

For quality improvements, we've seen that the audit log on investigating why quality issues are happening can be directly extracted from the video system. We typically think of using the video system for investigating theft, but if you have a defect happening on the shop floor or damage occurring during transportation, the video surveillance system is a resource to help you uncover the event that created the problem.

We've supported a large steel manufacturer that watches each roll of steel move through the facility. They have a clear audit of everywhere that it sat for how long, and they're collecting that metadata to know where things have gone wrong. A roll of steel is a very valuable asset and making sure that the quality is right is essential to this business's reputation. They've leveraged video analytics to track that product through their floor and know who touched it, when they touched it, and how it was touched to ensure that the quality going out to the customer is high, thereby improving profitability.

INCREASING EFFICIENCY

Efficiency improvements are a topic that's close to my heart. During my time at Boeing, I earned a Six Sigma Black Belt, which is all about identifying and eliminating inefficiencies through careful observation and time studies. Back then, it often meant spending hours on the shop floor with a clipboard, analyzing processes to pinpoint areas for improvement.

In the security space, I've seen how video analytics can do the same work—only faster and more accurately. It can identify wasted motion, accumulated inventory, or over-processing in real-time, all without needing someone to walk around with a clipboard. Instead, you get precise images and data that reveal exactly where inefficiencies are lurking. For a lean process improvement expert, this kind of actionable, visual data is a game-changer. It turns what used to be a manual, time-consuming task into a streamlined process that helps drive real improvements across the factory floor.

TURNING SECURITY DATA INTO GOLD

We've explored many ways your security system can add value and generate revenue. It's worth pausing to rethink its role within your organization. Instead of seeing it as just a cost, consider how it can be an efficiency enabler and a profit driver. Take the retail example—if shifting product placement boosts sales by $10,000 a month, that extra revenue can more than cover the cost of your security system. It pays to reach out to different departments and ask, "Here's the data we have—how can it help you? How can we use it to benefit the company?" You might be surprised at how many opportunities are waiting to be uncovered.

One particularly interesting example I worked on involved a hospital system with high-cost resources. Traditionally, each machine required a dedicated staff member to monitor it, even when it wasn't in use. This meant the hospital needed as many operators as machines, but they started to rethink that model.

Using their cameras to monitor these assets remotely, they asked, "Could one person oversee multiple machines from a central location?"

It reminded me of the evolution from piloted fighter jets to unmanned aircraft operated remotely. It's not about one person per machine; it's about managing more efficiently and increasing productivity.

This shift in mindset can transform how your security system supports your business. It's all about looking beyond the traditional and finding new ways to leverage the tools you already have.

UNLOCK NEW VALUE FOR YOUR BUSINESS

In this chapter, we explored that your security system's data isn't just about keeping the building safe—it's about unlocking a whole new level of value for your business. There's a treasure trove of insights hiding in that footage and access data, and when you dig in, you can turn it into something much more than just security. Whether it's improving customer experiences in retail, refining operations on the shop floor, or helping your team work more efficiently, these insights can become real game-changers for your organization.

But with all this potential comes a need for responsibility. It's not just about collecting and using data—you have to handle it carefully. That means making sure you're following company policies, respecting privacy, and keeping everything compliant with the latest regulations. Just because you can collect a certain piece of data doesn't always mean you should. Keeping that balance is key to turning your security data into a true asset.

As you look to get buy-in from executives, this is where you can really make your case. Showing that a security system isn't just a cost, but a source of valuable insights that can help the whole business, makes for a strong argument. In the next chapter, we'll dig into the executive playbook—how to make a compelling case for investing in security that goes beyond just safety. Knowing how to highlight the value in your data can be the key to getting the support you need.

CHAPTER 9

THE EXECUTIVE'S PLAYBOOK— CRAFTING A COMPELLING SECURITY INVESTMENT CASE

et me tell you about a security executive I work with. He's a brilliant physical security expert with a deep background in law enforcement and protecting spaces. One of the toughest parts of his job is convincing other executives that the security system is necessary. When the system reaches its end of life, he has to persuade them again to invest in a refresh. In the traditional approach, he's constantly having to resell or reconvince the organization to implement or update the security system.

Meanwhile, some IT executives have taken a different path. Instead of relying on the traditional capital expenditure model with large, one-time purchases, they've switched to leasing equipment. They know that every five years, the equipment will get refreshed because that's the agreement with their vendor. When the hardware reaches the end of its useful life, it's replaced with new gear, and they enter a new five-year agreement.

That security executive—despite all his skills and expertise—ends up spending a lot of time trying to get the organization on board with system upgrades. A more effective approach would be to change the way the organization invests in its security system, shifting from a capital expenditure model to an operating expense or lease agreement. This way, the hardware stays up to date without constant negotiation.

This chapter explores the shift from traditional capital expenditure models to modern operational expenditure approaches in security investments. It shows how this transition can lead to a more agile, cost-effective, and robust security setup. Let's start by defining the differences between these two approaches.

TWO MAIN MODELS: TRADITIONAL PURCHASE AND MODERN LEASE

In the world of security system investments, two main approaches define how organizations manage their expenses: the traditional purchase model and the "hardware as a service" model.

With the traditional purchase model, also known as capital expenditure (CapEx), organizations make a significant upfront investment to buy all the necessary equipment—cameras, card access systems, servers, and more. Once the equipment is installed, it is depreciated over time as part of the organization's accounting practices.

On the other hand, the hardware-as-a-service model takes a different route. Here, the service provider retains ownership of the equipment and takes on the responsibility of maintaining it throughout its lifespan. If a component fails, they handle the replacement, and when the system reaches the end of its life, they swap it for new technology under a renewed lease agreement.

Both of these models come with their own set of advantages and challenges. They offer different ways to finance and maintain a robust security system.

COMPARISON OF CAPEX VS OPEX FINANCING FOR SECURITY SYSTEMS

	PROS	CONS
CAPEX	Allows finance to capitalize asset on balance sheet and depreciate over time	Large upfront cash used to purchase equipment --- Replacement lifecycles are often not budgeted for
HARDWARE AS A SERVICE	Lower up-front cash use --- End of life replacements built into the contract	Capital assets not shown on the balance sheet and depreciated

When it comes to investing in a security system, businesses typically have two financing options: Capital Expenditure (CAPEX) or Hardware as a Service (HaaS). Each approach has distinct advantages and trade-offs that impact financial planning, budgeting, and long-term security system reliability.

Let's break it down:

CAPEX, or Capital Expenditure, is the traditional model where a company purchases security equipment outright. One of the main advantages is that these assets are recorded on the balance sheet and depreciated over time, which can have tax benefits. However, the downside is that CAPEX requires a significant upfront cash investment. Additionally, many organizations fail to budget for future replacement cycles, leading to outdated equipment that may eventually become a liability.

On the other hand, Hardware as a Service (HaaS) shifts security infrastructure costs into an operational expense model. Instead of purchasing hardware outright, companies lease the equipment through a service contract, making it easier to manage cash flow. This model ensures that end-of-life replacements are built into the agreement, eliminating the risk of security systems becoming obsolete. However, since the business does not own the hardware outright, these capital assets do not appear on the company's balance sheet, which may impact financial reporting in certain industries.

Both approaches have their place, depending on your organization's financial strategy and long-term security goals. If ownership and depreciation benefits are a priority, CAPEX may be the right fit. If predictable costs and automatic upgrades are more important, HaaS offers a streamlined alternative.

Ultimately, the choice between these models comes down to how your business balances upfront investment, long-term planning, and operational efficiency.

There's no one-size-fits-all solution. A capital expenditure model can work just as well, provided there's a solid plan for budgeting replacements over time. In this chapter, we're going to explore how these two approaches—CapEx and leasing—are perceived by executives, and what factors might make one approach more appealing than the other.

The following graph shows the lifecycle of hardware as a service versus CapEx.

CAPEX VS. HARDWARE-AS-A-SERVICE (HAAS):
Comparing Two Financing Models for Security Systems

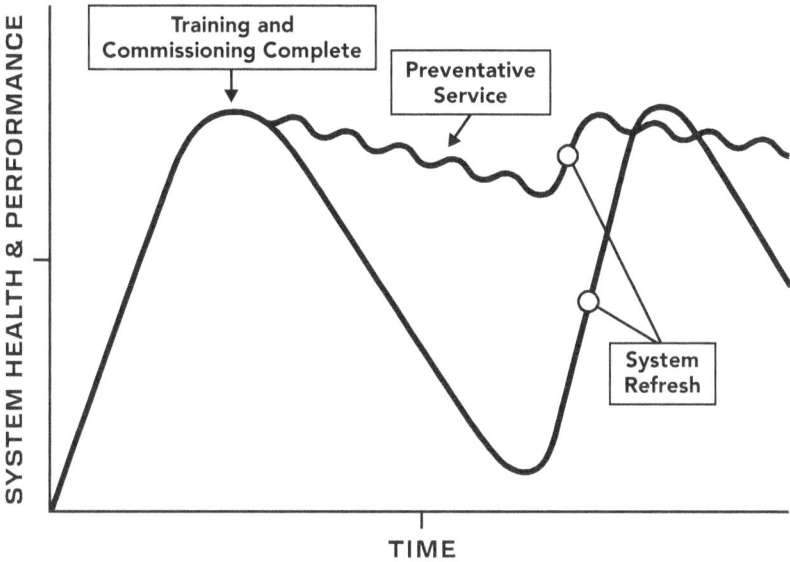

A visual comparison of traditional capital expenditure (CapEx) purchasing versus Hardware-as-a-service (HaaS), highlighting the performance and cost implications of each model over time.

This graphic illustrates how security system performance plays out over time under two different financing models: traditional capital expenditure (CapEx) versus Hardware-as-a-Service (HaaS), which is an operating expense (OpEx) approach.

The CapEx model is like buying a house—you pay upfront, own the equipment, and are responsible for repairs and replacements. In the graphic, we see system performance start strong, then decline steadily until there's a big dip when the system reaches end-of-life. At that point, we have to make a major reinvestment to bring performance back up. That "dip and restart" pattern repeats every few years.

With HaaS, it's more like renting. We own and maintain the hardware, and we proactively replace components as they age. While there's still some natural performance decline, we refresh the system before it tanks—keeping the performance curve much more stable and predictable. That's the difference here: proactive lifecycle management versus reactive replacement.

We'll look at the strategic considerations behind each model and how they align with an organization's financial and operational priorities.

THE FINANCIAL IMPLICATIONS

It's important to recognize that both of these approaches come with their own financial implications. If this is something you're considering, set up a meeting with your finance department to understand your organization's preferences around expenditures and cost management. The key focus here is on effectively spreading costs over the system's lifecycle and planning for replacements when needed. Some finance teams lean away from the cash flow strain of a large capital expenditure, as it requires a significant lump sum investment.

There are also tax implications to consider, such as how deductions are handled. Capital expenditures follow different depreciation schedules, affecting how and when you can recognize those tax benefits. This is a complex area, and it's best to defer to your finance team for detailed guidance.

Lastly, don't overlook the cost of system maintenance, especially if you don't have an extended warranty on capital expenditures. An extended warranty can cover replacement costs if equipment fails. In an operating expense model, any failure during the service agreement's life means the service provider is responsible for replacing the equipment. But with a capital expenditure approach, you might be on the hook to buy and install new equipment if something breaks. When comparing the financial aspects of these models, factor in maintenance and warranty costs to ensure a true apples-to-apples comparison.

TECHNOLOGICAL ADVANTAGES

Now, let's talk about some real technological perks of the operating expense model. Dealing with obsolete systems can be a headache, as we covered in earlier chapters. If you're stuck without the budget to replace your system, you'll miss out on those new features and updates that come with an operating model. That's a big win for the operating expense route—it keeps your tech up-to-date without scrambling for big bucks.

Another great thing about this model is how easy it is to scale. Need to add a few more cameras? With an operating expense model, those additions don't require a big lump sum; they're just small tweaks to your budget. It's easier than hunting down large amounts of money every time you need to expand.

And, let's not forget about avoiding obsolescence.

We worked with a library system that used the traditional model. They bought new systems when renovating branches, but those renovations happened every 10 to 15 years—way longer than the gear's useful life, which is more like seven years. So, they were stuck with outdated systems, without the budget to upgrade between renovation cycles.

Tired of this cycle, they switched to an operating expense model with a lease renewal every seven years. This made budgeting a breeze and ensured regular equipment refreshes. The branch managers loved it—they always had the latest tech that worked and supported their needs, without the hassle of dealing with outdated equipment.

OPERATIONAL BENEFITS

Let's talk about one last perk of the operating or lease subscription model: the operational benefits. In the traditional setup, the owner is on the hook for everything—keeping the system running, maintaining it, and troubleshooting issues. If you bought it, it's your problem to fix. But with the operating model, that responsibility shifts from you to a trusted vendor who's accountable for making sure everything works—keeping cameras running, ensuring the card access system functions smoothly. It takes a lot of the operational burden off your plate.

Plus, when that responsibility shifts to an outside vendor and they know it's their contractual duty, they take it seriously. They have a vested interest in keeping the system in top shape. Another advantage is that this model aligns better with how most businesses are managing their IT systems these days. We've seen the shift from on-premises servers to cloud-based ones, from buying IT systems outright to leasing them, become the norm—with over 50% of commercial systems now purchased this way. By making this shift, you're aligning your organization more closely with modern IT practices.

MODERNIZE YOUR SECURITY STRATEGY

From my experience, transitioning from the traditional CapEx model to an operating model is a comprehensive step toward modernizing your security strategy. This shift helps keep your system maintained, ensures timely replacements before obsolescence sets in, and aligns with IT standards to reduce risks. It's a move that many executives in your organization will appreciate, as it offers them a predictable and manageable way to handle budgeting, steering clear of unexpected expenditures.

To explore this shift, start the conversation with your finance department. Discuss the potential benefits of moving away from large, one-time capital expenses and towards a more consistent operating expense. Emphasize the advantages of better cash flow management and smoother budget planning, and see if they would support a transition from a CapEx model with depreciation to an ongoing, steady expense.

Often, organizations are more open to this change than you might expect. After that discussion, reach out to your current vendor—or consider contacting us—to talk about how to implement this shift. At IP Systems, we've adopted this approach in regulated spaces because it provides a reliable, ever-current system with full support throughout its lifecycle. We've made it our standard offering, and your existing vendor might be able to do something similar. Ask about a five-to-seven-year equipment lifecycle lease agreement that ensures your system stays up to date.

Keep in mind that many vendors still follow the traditional model. This operating model draws more from IT practices, but we see it as the future of security solutions. Don't be surprised if your vendor hasn't caught up yet. If that's the case, you're welcome to reach out to us at IPsystems.tech. Visit our website, fill out the form, and let us know you're interested in security as a service and making the shift to an operating model. We'll have someone from our team get in touch with you promptly.

CONCLUSION: THE ROAD AHEAD TO SECURITY MASTERY

So here we are.

If there's one thing I hope this book has made clear, it's that the world of security is overdue for a shift from the dusty old phone book model to something smarter, faster, and more future-proof.

We don't rely on outdated tools in any other part of our lives. We don't pull out the phone book when we need directions or look up a neighbor's address. So why would we tolerate outdated systems when it comes to protecting the people, assets, and spaces that matter most?

The technology is here. The tools exist. But the way we think about security—the way we buy it, maintain it, and plan for it—needs to evolve. That's what this book is about: pushing for a new era of security, one where systems are maintained, upgraded, and aligned with the speed of modern threats.

In today's world, a reactive security strategy is as outdated as a phone book. Relying on either is a risk you can't afford to take.

And speaking of risk, I can't express how grateful I was that the security system at the business where I witnessed the double homicide was working. It became crucial evidence in identifying and convicting the murderer. The importance of having a reliable and up-to-date system that protects your business and loved ones has driven me ever since. It's fueled my passion since 2017, leading us to deploy over 20,000 continuously monitored, modern devices. We have the confidence to tell our customers that their systems are working, and if they're not, we're taking immediate action to resolve any issues. This commitment to ensuring the system is always operational gives me peace of mind. I hope it does the same for you, knowing that the modern security approach is not only more reliable but also better supported. It's what you need to safeguard your family and your business.

PUSH SECURITY FORWARD

I wrote this book because the security industry is stuck in the past. For all the innovation we've seen in technology—facial recognition, AI-powered analytics, biometric scanning—how security is delivered and maintained hasn't caught up. It's like we're living in a world of self-driving cars, but still relying on payphones and rotary dials to call for help. The systems themselves are evolving, but the processes surrounding them remain static.

Worse, I've seen highly capable, well-intentioned security executives held back by the outdated mindset of how security is purchased, managed, and maintained. They want to do more, but the current model gets in their way.

This book is my attempt to push the conversation forward. I want to lay out a better way—one rooted in proactive support, intelligent maintenance, and real-time monitoring. When we stop treating security like a one-time purchase and start treating it like the mission-critical system it is, the results are game-changing. People are safer. Assets are more secure. And organizations run smoother.

THE REAL DIFFERENCE OF A MODERN SECURITY APPROACH

If you've made it this far, then you're already ahead of the curve. You understand that security can't be static. It has to adapt, evolve, and grow with your organization. The question is: what are you going to do about it?

My hope is that this book has given you not just the inspiration, but the tools to take action. Whether that's reevaluating your current system, exploring a more service-based model, or simply building a better maintenance plan. Whatever your next step is, don't wait. The pace of change is accelerating, and the longer you put it off, the harder it will be to catch up.

I'm excited for the journey you're about to embark on. This book has taken you through every aspect of modern security management. Embracing the modern security approach will make a meaningful impact for you.

Let's revisit what we've explored together. We started by highlighting the risks of sticking with an outdated system—how new threats keep emerging and how unchecked vulnerabilities can put you at risk. It was key to equip you with the knowledge that implementing a security system is not a solo effort. It's about building a strong team of experts to guide and support you along the way.

Then, we delved into the SAFE approach to modern security. This began with conducting surveys to assess the specific risks facing your business and setting clear guidelines to create an unbreakable security plan. From there, we moved on to the practical steps of applying and implementing that plan with the help of trusted security experts. We also discussed how to fortify your system—keeping it maintained, updated, inspected, and performing at its best. We stressed the need for your system to evolve, staying current and effective as new challenges arise.

We also uncovered the hidden potential in the data your security system generates. This includes how you can transform that data into a strategic asset that underscores the value of your security investment. We explored the shift from the traditional capital expenditure model to a more consistent and manageable security-as-a-service approach, ensuring your system remains up to date through leasing.

With all these insights, you're now ready to make thoughtful, strategic choices that will shape a safer, smarter future for your organization.

You don't have to do this alone. My team and I are here to help guide you through the transition—from wherever you are today to where you need to be tomorrow. As you work to implement the Modern Security system, think of us as your partner and resource. Visit our website at www.ipsystems.tech to explore a variety of tools and resources designed to support you every step of the way.

If you need any assistance, reach out to see how we can help. This is a big undertaking, but there are plenty of resources to guide you through the process.

Let's modernize security together.

ACKNOWLEDGMENTS

I would love to share the names of companies and individuals I have been able to work with, but the security industry likes to be anonymous and discreet. So I will stick to the first names of the people that I respect, appreciate, and call friends: David, Darrin, Tony, Melisa, Brandon, Dean, Nelson, Leland, Roger, Jake, and Spencer. I value each of you immensely and thank you for your friendship.

I also want to thank Jim Butkovic and Greg Ponchak who trusted me with IPS when they retired. To Ken, Valerie, Tyler, Eric, Ron, and Mike, thank you for joining the FOLD and being instrumental in the growth of IPS.

To my wife Beth for supporting and loving me as we had 3 incredible kids in 33 months, while I bought a business and learned about the security industry.

ABOUT THE AUTHOR

Rob Jackson is the President of Integrated Precision Systems (IPS), a national security technology company that designs, deploys, and manages modern security infrastructure for some of the most complex and high-risk environments in the country—including correctional facilities, critical infrastructure sites, and nuclear power plants. Under his leadership, IPS has expanded to serve clients in 27 states, monitoring over 20,000 devices every 60 seconds to ensure safety, compliance, and peace of mind.

Rob holds an MBA from the MIT Sloan School of Management, where he focused on technology strategy and entrepreneurship, and a BS in Mechanical Engineering from the University of Southern California. His career began in aerospace as a production engineer on the 787 Dreamliner program, where he developed a deep respect for mission-critical systems. He later joined McKinsey & Company, advising Fortune 500 clients on operations and business transformation, and went on to co-found a tech-enabled laundry and logistics startup in New York City.

What drives Rob today is the urgent need to bring the security industry into the 21st century. While technology has evolved rapidly, the way organizations buy, maintain, and manage security systems has remained stagnant. Rob wrote this book to challenge the outdated paradigms of security management and to equip security leaders with the tools and mindset needed for proactive, technology-driven protection.

When he's not building systems that keep people and property safe, Rob is likely off the grid—summit climbing, scuba diving, skiing, or spending time with his family on a boat.

APPENDIX I: MAINTENANCE CHECKLIST FOR MODERN SECURITY SYSTEMS

Need help with maintenance for your modern security system? These checklists will help you keep your system maintained, daily, weekly, monthly, and annually.

DAILY MAINTENANCE

1. Camera System Checks: ☑

   a. Test all camera feeds for clear visuals and connectivity.

   b. Verify that recording and playback functions are working seamlessly.

   c. Ensure that all cameras are aligned correctly and cover critical areas.

2. Access Control System Review: ☑

   a. Monitor and review access logs for unusual or unauthorized activity.

   b. Confirm that biometric scanners, keycard readers, and digital locks function without delays.

   c. Inspect doors, gates, and windows for any physical damage or signs of tampering.

3. Environmental Controls: ☑

   a. Check that lighting systems around critical areas (e.g., entrances, exits, perimeter) are fully operational.

   b. Ensure temperature and humidity levels in server and equipment rooms are within optimal ranges to avoid system failures.

4. Immediate Threat Response Readiness: ☑

   a. Confirm all emergency communication lines (e.g., intercoms, alert buttons) are operational.

   b. Conduct quick spot checks on security personnel for readiness and alertness.

WEEKLY MAINTENANCE

5. Surveillance Systems Diagnostics: ☑

   a. Run diagnostic tests on all surveillance software to detect potential issues.

   b. Inspect camera lenses for dust, dirt, or obstructions; clean as needed.

6. Alarm and Intrusion Systems: ☑

   a. Test all alarm systems (including fire, intrusion, and panic alarms) for responsiveness.

   b. Validate that motion detectors, glass break sensors, and door/window sensors are functioning as intended.

7. Perimeter and Barrier Inspections: ☑

   a. Check the integrity of fences, gates, and barriers for potential vulnerabilities.

   b. Ensure external lighting and motion-activated systems cover all necessary areas effectively.

8. Access Control Maintenance: ☑

   a. Review staff and visitor access permissions to ensure they are up to date.

   b. Verify that backup power sources (e.g., batteries, generators) are operational and ready.

MONTHLY MAINTENANCE

9. Cybersecurity Protocol Updates: ☑

   a. Update passwords, encryption keys, and access credentials across all systems.

   b. Run full malware scans and ensure firewalls and antivirus software are updated to the latest versions.

10. Hardware and System Integrity Checks: ☑

   a. Inspect sensors, keypads, control panels, and related hardware for wear or defects.

   b. Replace any aging batteries in wireless sensors, alarms, and communication devices.

11. Personnel and Training Assessments: ☑

   a. Review incident response protocols and conduct refresher training sessions for staff.

   b. Evaluate recent security events or near misses and adjust procedures accordingly.

12. Data Storage and Backup Review: ☑

   a. Ensure that video footage, access logs, and system backups are securely stored and easily retrievable.

   b. Confirm that cloud storage or local backup systems are functioning correctly.

ANNUAL MAINTENANCE

13. Comprehensive System Audit and Threat Assessment: ☑

   a. Perform a full audit of all hardware, software, and procedures to identify vulnerabilities.

b. Conduct a professional threat assessment to update risk profiles and response plans.

14. Upgrades and Professional Inspections: ☑

a. Schedule certified professionals to inspect and calibrate cameras, sensors, access control devices, and alarms.

b. Evaluate the need for system upgrades or integrations with new technologies (e.g., AI analytics, cloud management).

15. Regulatory Compliance Review: ☑

a. Ensure all security measures are in line with local, state, and federal regulations.

b. Review licenses, certifications, and training records to maintain compliance with legal and industry standards.

16. Disaster Recovery Plan Update: ☑

a. Review and update the disaster recovery and business continuity plans.

b. Conduct drills and tests for emergency scenarios like power outages, natural disasters, or cyberattacks.

APPENDIX II: SECURITY BEST PRACTICES CHECKLIST

This checklist can help you implement effective, easy-to-follow best practices for securing your facilities.

⊘ PHYSICAL SECURITY

- Regularly inspect and reinforce all doors, windows, and barriers.
- Ensure all areas have adequate lighting, particularly at night.
- Limit access to sensitive areas using keycards, PINs, or biometrics.

⊘ CYBERSECURITY

- Implement multi-factor authentication for all system logins.
- Regularly update firewalls, antivirus software, and encryption protocols.
- Train staff on cybersecurity hygiene, including password management and phishing awareness.

⊘ PERSONNEL TRAINING

- Conduct routine security drills, including fire, lockdown, and evacuation exercises.
- Update staff on emerging threats and changes in protocols.
- Use role-specific training to ensure every employee knows their responsibilities in a crisis.

APPENDIX III: COST CONSIDERATIONS FOR UPGRADING OR MAINTAINING SECURITY SYSTEMS

When planning for security, it's essential to weigh the upfront and ongoing costs of upgrading versus maintaining your existing systems. Here's an overview to help set expectations:

☑ UPGRADING COSTS:

1. Initial Investment:

a. Purchasing cutting-edge hardware such as AI-enhanced cameras, biometric scanners, and advanced sensor systems.

b. Software licenses and subscriptions for cloud-based management, threat detection, and automated monitoring solutions.

2. Installation and Integration:

a. Professional fees for configuring and integrating new systems with existing infrastructure.

b. Training for staff on operating upgraded systems, including any new protocols.

3. Long-Term Value:

a. Enhanced security can lead to lower insurance premiums and reduced vulnerability to costly breaches.

b. Advanced systems may offer automation features, reducing the need for extensive manual monitoring.

☑ MAINTAINING EXISTING SYSTEMS:

4. Recurring Costs:

a. Ongoing repairs for aging hardware like cameras, access panels, and sensors.

b. Regular software updates, maintenance contracts, and personnel training.

5. Risk and Vulnerability:

a. Delaying upgrades can leave gaps in coverage and expose your facility to evolving threats.

b. Older systems may struggle to integrate with newer technology, leading to inefficiencies.

6. Long-Term Expenses:

a. Accumulated costs from frequent repairs and emergency fixes could surpass the cost of an upgrade over time.

b. Potential liability from outdated systems failing to detect breaches or respond effectively.

c. Understanding these costs will help you make informed decisions that align with your budget, security goals, and long-term strategy.

APPENDIX IV: ENRICHED SECTIONS FOR SECURITY PROFESSIONALS

GLOSSARY OF KEY TERMS

- **ACCESS CONTROL:** The selective restriction of access to a place or resource, typically through keycards, biometric systems, or digital passwords.

- **BIOMETRIC AUTHENTICATION:** Security processes using unique biological traits like fingerprints, facial recognition, or iris scans to verify identity.

- **INTRUSION DETECTION SYSTEM (IDS):** A system that monitors network traffic for suspicious activity and potential breaches.

- **MULTI-FACTOR AUTHENTICATION (MFA):** A layered approach requiring multiple forms of verification before granting access.

- **REDUNDANCY PROTOCOLS:** Backup systems and strategies designed to ensure continued operations if primary systems fail.

- **ZERO-TRUST ARCHITECTURE:** A security model where no entity, inside or outside the network, is trusted by default and must be verified continuously.

RESOURCE LIST

This resource list can help you easily find trusted suppliers and further guidance when planning or upgrading your security systems.

Industry Organizations and Standards:

- **ASIS INTERNATIONAL:** Provides globally recognized certifications like the CPP (Certified Protection Professional).

- **ISC²**: Offers certifications such as CISSP (Certified Information Systems Security Professional), vital for cybersecurity.

Recommended Vendors:

- **ACCESS CONTROL**:
  - Honeywell (honeywell.com)
  - HID Global (hidglobal.com)
  - Genetec (genetec.com)
- **SURVEILLANCE SYSTEMS**:
  - Axis Communications (axis.com)
  - Milestone (milestonesys.com)
  - Avigilon (avigilon.com)
  - Hanwha Vision (hanwhavisionamerica.com)

Further Reading:

- "Security Management Standard: Physical Asset Protection" (ASIS International)
- "NIST Cybersecurity Framework" (National Institute of Standards and Technology)

COMMON THREAT SCENARIOS AND RESPONSE PROTOCOLS

These scenarios will help prepare you for real-life threats, offering clear steps to take in each situation.

- **SCENARIO**: Unauthorized Access Attempt
  - **DETECTION**: An employee badge is used after hours in a restricted area.
  - **IMMEDIATE RESPONSE**: Security alerts the control room; access logs are reviewed; guards are dispatched to the location.
  - **FOLLOW-UP**: The badgeholder is questioned, and access privileges are temporarily revoked pending investigation.
- **SCENARIO**: Power Outage
  - **DETECTION**: A sudden blackout disrupts surveillance and access control systems.
  - **IMMEDIATE RESPONSE**: Backup generators power up critical systems; staff are positioned at key entry points.
  - **FOLLOW-UP**: Investigate the cause, review emergency power protocols, and perform drills to improve readiness.

TIMELINE OF TECHNOLOGICAL ADVANCEMENTS IN SECURITY

This timeline gives you a quick overview of how security technology has evolved, helping you understand where current systems fit into the broader landscape.

- **1980S: ANALOG CCTV SYSTEMS**: The standard for video surveillance, offering real-time monitoring but limited storage and low resolution.
- **1990S: DIGITAL VIDEO RECORDING (DVR)**: Improved image

quality and storage capabilities, making footage more accessible.

- **2000S: IP-BASED SURVEILLANCE:** Introduction of internet-connected cameras with remote access and high-definition quality.

- **2010S: AI AND ANALYTICS INTEGRATION:** Introduction of facial recognition, behavior analysis, and real-time alerts.

- **2020S: CLOUD AND HYBRID SECURITY MODELS:** Cloud storage and AI-enhanced systems become mainstream, offering scalable and flexible solutions.

CASE STUDIES OF SECURITY FAILURES AND LESSONS LEARNED

These case studies offer insights into real-world failures and practical steps to avoid similar pitfalls.

- **CASE STUDY:** Data Breach Due to Outdated Software

  - **INCIDENT:** A major retail chain suffered a breach when hackers exploited vulnerabilities in an outdated system.

  - **KEY LESSON:** Regular software updates and proactive threat assessments could have prevented the breach.

- **CASE STUDY:** Inadequate Perimeter Security at a High-Security Facility

  - **INCIDENT:** An inmate escaped due to insufficient perimeter monitoring and delayed response.

  - **KEY LESSON:** Ensure multiple layers of monitoring, including motion detectors, cameras, and manned patrols, are in place.

VENDOR COMPARISON CHART

This chart is an example of how you can compare the features and capabilities of different security solutions in your business. Once you know your security standards, you can use a table like this to evaluate and select the solution that is best for your organization.

VENDOR COMPARISON CHART

FEATURE	VENDOR A	VENDOR B	VENDOR C
ACCESS CONTROL	INTEGRATED BIOMETRICS, KEYCARD OPTIONS	BIOMETRIC ONLY	KEYCARD ONLY
VIDEO SURVEILLANCE	4K RESOLUTION, AI ANALYTICS	HD RESOLUTION, BASIC ANALYTICS	1080P, NO ANALYTICS
SCALABILITY	ENTERPRISE-LEVEL, SUPPORTS MULTI-SITE	SUITABLE FOR MID-SIZE BUSINESSES	BEST FOR SMALL FACILITIES
SUPPORT AND WARRANTY	24/7 SUPPORT, 5-YEAR WARRANTY	LIMITED SUPPORT, 3-YEAR WARRANTY	9-5 SUPPORT, 1-YEAR WARRANTY

Hardware replacement timelines for common security system components, based on typical usage.

APPENDIX V: THE COPENHAGEN LETTER

Copenhagen, 2017

To everyone who shapes technology today:

We live in a world where technology is consuming society, ethics, and our core existence.

It is time to take responsibility for the world we are creating. Time to put humans before business. Time to replace the empty rhetoric of "building a better world" with a commitment to real action. It is time to organize, and to hold each other accountable.

Tech is not above us. It should be governed by all of us, by our democratic institutions. It should play by the rules of our societies. It should serve our needs, both individual and collective, as much as our wants.

Progress is more than innovation. We are builders at heart. Let us create a new Renaissance. We will open and nourish honest public conversation about the power of technology. We are ready to serve our societies. We will apply the means at our disposal to move our societies and their institutions forward.

Let us build from trust. Let us build for true transparency. We need digital citizens, not mere consumers. We all depend on transparency to understand how technology shapes us, which data we share, and who has access to it. Treating each other as commodities from which to extract maximum economic value is bad, not only for society as a complex, interconnected whole but for each and every one of us.

Design is open to scrutiny. We must encourage a continuous, public, and critical reflection on our definition of success as it defines how we build and design for others. We must seek to design with those for whom we are designing. We will not tolerate design for addiction, deception, or control. We must design tools that we would love our loved ones to use. We must question our intent and listen to our hearts.

Let us move from human-centered design to humanity-centered design.

We are a community that exerts great influence. We must protect and nurture the potential to do good with it. We must do this with attention to inequality, with humility, and with love. In the end, our reward will be to know that we have done everything in our power to leave our garden patch a little greener than we found it.

We who have signed this letter will hold ourselves and each other accountable for putting these ideas into practice. That is our commitment.